New Perspectives and Applications of Modern Control Theory in Engineering

Edited by Constantin Volosencu

Published in London, United Kingdom

New Perspectives and Applications of Modern Control Theory in Engineering
http://dx.doi.org/10.5772/intechopen.1008578
Edited by Constantin Volosencu

Contributors
Constantin Volosencu, Fatiha Zidani, Kok Hwa Lim, Tan Yong Rui, Victor Martinez-Luaces, Victor Siang Tze Sim

First published in London, United Kingdom, 2026 by IntechOpen
IntechOpen is the global imprint of INTECHOPEN LIMITED, registered in England and Wales, registration number: 11086078, 167-169 Great Portland Street, London, W1W 5PF, United Kingdom

For EU product safety concerns: IN TECH d.o.o., Prolaz Marije Krucifikse Kozulić 3, 51000 Rijeka, Croatia, info@intechopen.com or visit our website at intechopen.com.

British Library Cataloguing-in-Publication Data
A catalogue record for this book is available from the British Library

New Perspectives and Applications of Modern Control Theory in Engineering
Edited by Constantin Volosencu
p. cm.
Print ISBN 978-1-83635-454-3
Online ISBN 978-1-83635-453-6
eBook (PDF) ISBN 978-1-83635-455-0

Meet the editor

Constantin Voloşencu is a prominent Professor in the Department of Automation and Applied Informatics at the Politehnica University of Timişoara, Romania. His professional achievements are extensive, particularly in control systems, electrical drives, and intelligent systems (including fuzzy logic and neural networks). He is the author of over 10 books, 11 book chapters, and more than 170 scientific papers published in international journals and conference proceedings. He holds 27 patents and has managed numerous research grants. He has developed various electrical and automation systems for industrial applications, including machine tools and high-power ultrasound processes. He serves as editor-in-chief and a member of the editorial boards of international journals; he has served on scientific committees at many international conferences and has been a plenary and keynote speaker at numerous global conferences. Significant contributions to both academic research and practical engineering development characterize his career.

Contents

Preface

Control theory is a field in continuous evolution, driven by two primary factors: the emergence of novel mathematical methods successfully applied to control challenges, and the advancement of modern numerical computing equipment. This technological progress facilitates the implementation of complex computational algorithms that demand high processing speeds and substantial memory capacity. Modern control theory integrates methods from applied mathematics, computer science, and engineering. Based on this discipline, it is possible to understand the functioning of automated systems, their mathematical modeling, and, crucially, their design to achieve desired performance indicators. Following a remarkable evolution from classical linear systems theory, the field has progressed towards the study of nonlinear systems, intelligent systems, and complex interdisciplinary systems. The field of control theory attracts researchers to develop and innovate mathematical methods that find immediate practical application in modern engineering. By applying control theory, engineers can design systems that operate efficiently, meeting operational requirements. This theory has diverse applications across many sectors of contemporary society, from residential automation systems to aircraft, complex energy systems, and autonomous vehicles.

The book's motivation is to provide intuitive applications of rigorous theories, accessible for implementation in engineering practice, supported by case studies and examples that demonstrate the power and versatility of modern control methods.

This book is designed to bridge the gap between theory and practical application. The work presents examples and applications of fundamental concepts, including dynamical systems modeling, stability analysis using classical criteria (such as the Hurwitz criterion), and the design of PID controllers and state-feedback control systems. The volume also includes modern applications of robust fault-detection strategies in industrial motor drives, the application of control theory principles to solve inverse problems in chemical kinetics, and new methodologies for assessing cybersecurity risk in industrial control processes. Thus, the book is structured into five chapters, as follows. The chapters in this book are written by experts working in both academia and industry, each bringing unique perspectives on how contemporary control strategies are dynamically reshaping engineering fields. Chapter 1 serves as an introduction, detailing specific aspects of nonlinear control systems. Key topics covered include essential nonlinearities, the harmonic balance method, phase plane analysis (state portrait), Lyapunov's second stability theorem, and Popov's stability theorem. Each theoretical concept is thoroughly illustrated with calculated examples. This chapter aims to lay the groundwork by revisiting the fundamental concepts necessary for understanding systems where the output is not directly proportional to the input, a condition prevalent in most physical processes. Chapter 2 presents a sensor fault-tolerant control strategy specifically designed for induction motor drives. This strategy utilizes a field-oriented indirect control approach based on rotor flux orientation. The control structure employs a multi-observer system derived from Kubota observers. A comprehensive fault detection, isolation, and compensation process is integrated within the control scheme to maintain robust system stability and performance. Chapter 3 presents a study of a multitude of functions arising in first-order chemical kinetics problems, examining their inherent structure and related inverse problems. The existence, uniqueness, and stability of the solutions to these problems are rigorously analyzed. Chapter 4 proposes a methodology for assessing risks in cyber-physical processes. This approach integrates an analysis of protective layers with the quantification of cyber risks using the Factor Analysis of Information Risk (FAIR) framework. This methodology is applied to a practical case study involving a water treatment plant. Chapter 5 has an educational focus, presenting a synthesis of calculations for the stability domains of control systems. It explicitly addresses proportional-integral (PI) controllers for second- and third-order processes, using the Hurwitz stability criterion.

The applications and examples presented in the book will inspire readers and demonstrate the vast potential of control theory for solving today's engineering challenges. This book is intended for graduate students, faculty, researchers, and practicing engineers who wish to deepen their understanding of control theory and explore its modern implementations. Whether used as an educational textbook, a reference, or a source of inspiration for future innovations, we hope this work will contribute significantly to the ongoing dialogue between theory and practice in engineering control systems.

The editor would like to express its gratitude to the authors who contributed their excellent research work to the publication of this book. The editor would like to thank Mateo Belir, the Publishing Process Manager for the book, and all the members of the publishing house for their important contributions to the publication.

Constantin Voloşencu
Department of Automation and Applied Informatics,
Politehnica University Timisoara,
Timisoara, Romania

Chapter 1

Introductory Chapter: On Some Basic Concepts in Control Theory of Nonlinear Systems

Constantin Volosencu

1. General considerations

Nonlinear control theory applies to control systems in which the process or controller has complex structures and cannot be described by linear equations. The principle of superposition cannot be applied to nonlinear systems. Most processes in nature, industry or economics are nonlinear systems. However, they are analyzed based on the linear models obtained by initializing their primary physical models. To control these processes, nonlinear regulators are used in practice [1, 2]. Such nonlinear controllers are elements with saturation, hysteresis, or complex interactions between multiple components as fuzzy logic [3] or neural networks [4, 5]. Scientific works in the specialized literature have demonstrated that nonlinear regulators provide better performance indicators of control systems [3–5]. Nonlinear control systems are more robust to parameter estimation errors and to variations in disturbing variables [3–5]. The nonlinear control systems are using techniques that combine not only the concepts of linear regulation such as PID control suitable for systems with well-defined linear behavior but also specialized concepts for analysis and design. The complexity of analyzing and designing controllers for nonlinear systems is higher than for linear systems, and it is computationally intensive. Nonlinear systems are characterized by aspects that require specific treatment, different from linear systems.

This chapter presents some applications to some techniques of nonlinear control system analysis and design as: harmonic balance method, describing functions [3], phase plane analysis [4], Lyapunov theory [1, 2, 3, 6] and Popov's criterion [2]. These concepts are providing tools for analyzing the stability of nonlinear systems and designing controllers. Some calculation examples made with MATLAB are given.

2. Nonlinear control systems

The general block diagram of a nonlinear control system is presented in **Figure 1**.

In **Figure 1**, w is the reference variable, e is the regulation error, N is the nonlinear controller, u is the command variable, P is the linear process and r is the feedback variable. In simple control systems, essential nonlinear elements are used. The input-output characteristics of some nonlinear elements, such as (a) relay on-off, (b) relay with dead zone, (c) linear with saturation and (d) relay with hysteresis [6], are presented in **Figure 2**.

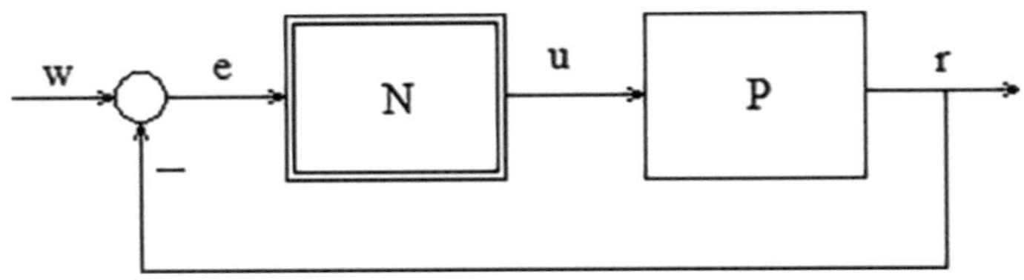

Figure 1.
The structure of a nonlinear system.

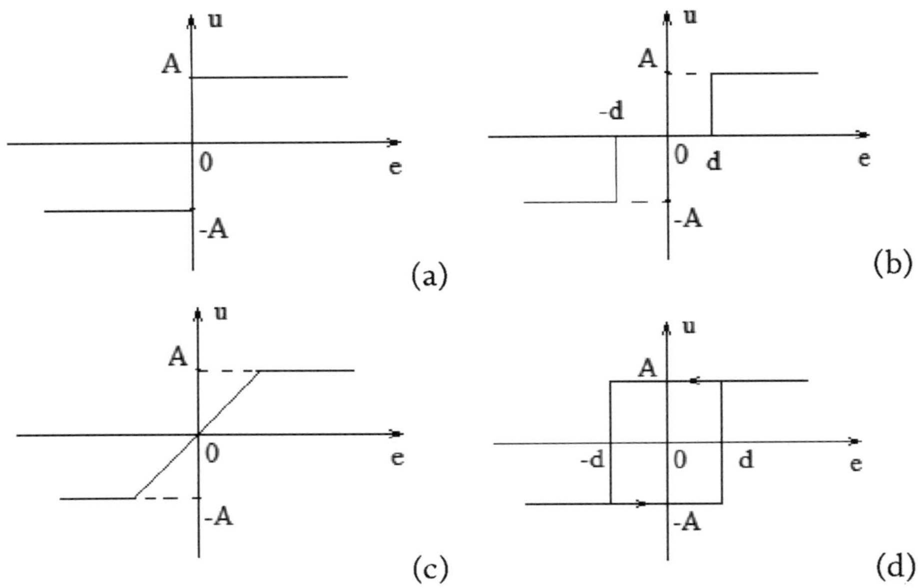

Figure 2.
Nonlinear elements.

Example. To understand how sustained oscillation occurs in nonlinear control systems, an example of a temperature control system often used in practice with a relay with hysteresis controller as in **Figure 2(d)** is analyzed. The process has a transfer function with delay time and an integrator

$$H_P(s) = \frac{1}{s} e^{-s\tau} \tag{1}$$

Time characteristics obtained in response to the step signal are shown in **Figure 3**.

It is observed that the system variables have a periodic character specific to *limit cycles* [6]. Sustained oscillations reduce the performance of control systems, reduce the reliability of systems, especially on the mechanical side, and cause inconvenience to users. That is why they need to be kept under control. In the analysis of limit cycles, $u = N(e(t))$ is a nonlinear function and the linear process has the transfer function

$$H(s) = \frac{B(s)}{A(s)} = \frac{b_m s^m + b_{m-1} s^{m-1} + \dots + b_1 s + b_0}{a_n s^n + a_{n-1} s^{n-1} + \dots + a_1 s + a_0} \tag{2}$$

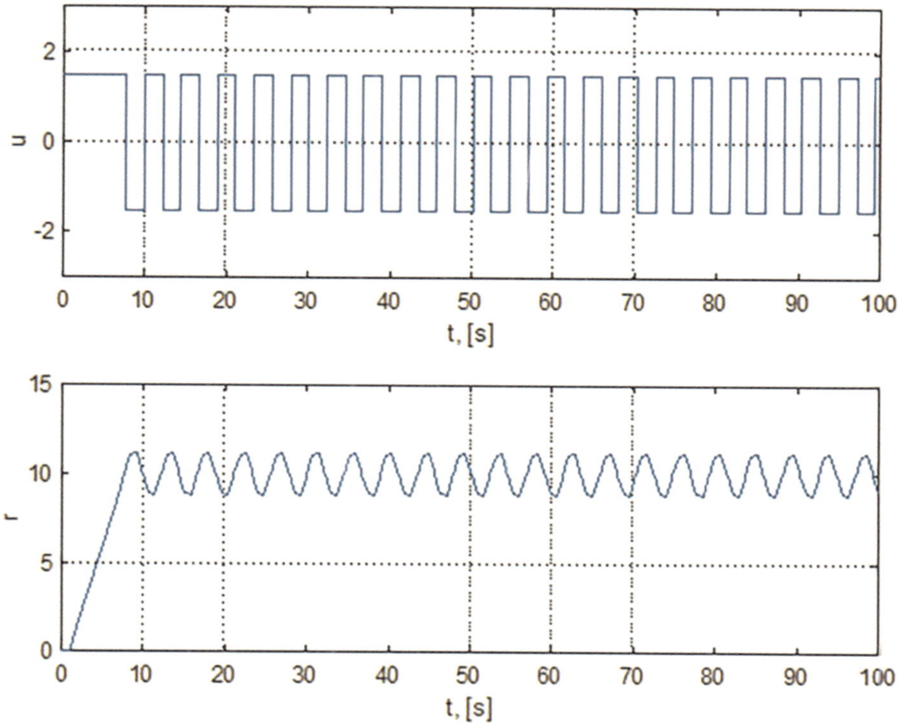

Figure 3.
Oscillatory characteristics.

The process P should be controllable and observable. The analysis is made for $w = 0$.

Harmonic balance method [7, 8]. An analysis in the frequency domain is required. The output and the input signals are represented in Fourier series

$$r(t) = \sum_{k=-\infty}^{+\infty} c_k e^{jk\omega t}, u(t) = \sum_{k=-\infty}^{+\infty} d_k e^{jk\omega t} \tag{3}$$

The equation of the harmonic balance is (4) and with Fourier expansion is (5)

$$A(j\omega)r(t) + B(j\omega).N(r(t)) = 0 \tag{4}$$

$$\sum_{k=-\infty}^{\infty} [A(kj\omega)c_k + B(kj\omega)d_k].e^{jk\omega t} = 0 \tag{5}$$

The function that describes the nonlinearity is called *describing function* [9]

$$\tilde{N} = \frac{d_1}{c_1} = \frac{2\omega}{\pi c}\int_0^{\frac{\pi}{\omega}} N(c\sin(\omega t))\sin(\omega t)dt = \frac{2}{\pi c}\int_0^{\pi} N(c\sin(\theta))\sin(\theta)d\theta \tag{6}$$

with $c = 2jc_1$ which is *the amplitude of the harmonic oscillations*.

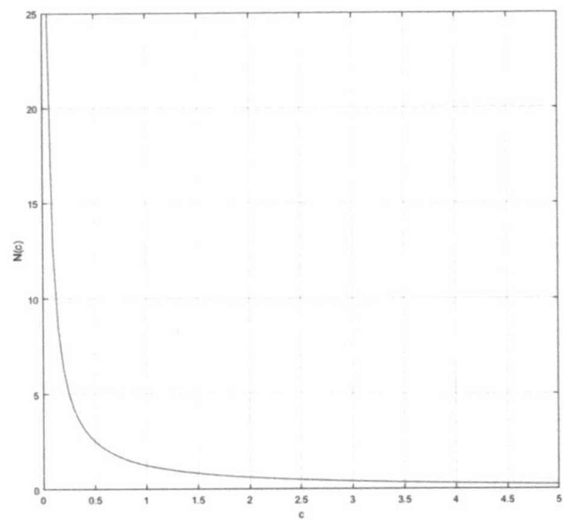

Figure 4.
Describing function graph.

Eq. (4) can be written as *the equation of the harmonic balance*

$$H(j\omega) = -1/\tilde{N}(c) \tag{7}$$

The solution of the equation of the harmonic balance: ω is determined at the intersections of hodograph $H(j\omega)$ with the real axis; c is determined at the intersections of hodograph $H(j\omega)$ with the locus of critical points $1/\tilde{N}(c)$.

Example. As an example of nonlinearity, a relay element is chosen, with the input-output relation (8) and the describing function (9).

$$N(e) = \begin{cases} U, & e > 0 \\ -U, & e < 0 \end{cases}, U = 1 \tag{8}$$

$$\tilde{N}(c) = \frac{2}{\pi c}\int_0^\pi N(c\sin\theta)\sin\theta . d\theta = \frac{2}{\pi c}\int_0^\pi U\sin\theta . d\theta = \frac{4U}{\pi c} \tag{9}$$

Its graphic is presented in **Figure 4**.

A third-order process is chosen, as a linear part, with the transfer function

$$H(s) = \frac{1}{(s+1)^3} \tag{10}$$

The hodograph of $H(j\omega)$ and the graph of $1/\tilde{N}(c)$ are presented in **Figure 5**.

The intersection points, which provide the parameters of the limit cycles, are highlighted. The intersection with the real axis is at -0.124. The amplitude of oscillation of r(t) is c = 0.123 x 4/π ≈ 0.56. Angular frequency of the intersection is ω = 1.74 rad/s. Period of the oscillations is T = 2π/ω ≈ 3.60 s.

Example. A time domain analysis of the above nonlinear control system is made in MATLAB. The graphs of the oscillating signals are shown in **Figure 6**.

The values of c and T are according to the above harmonic analysis.

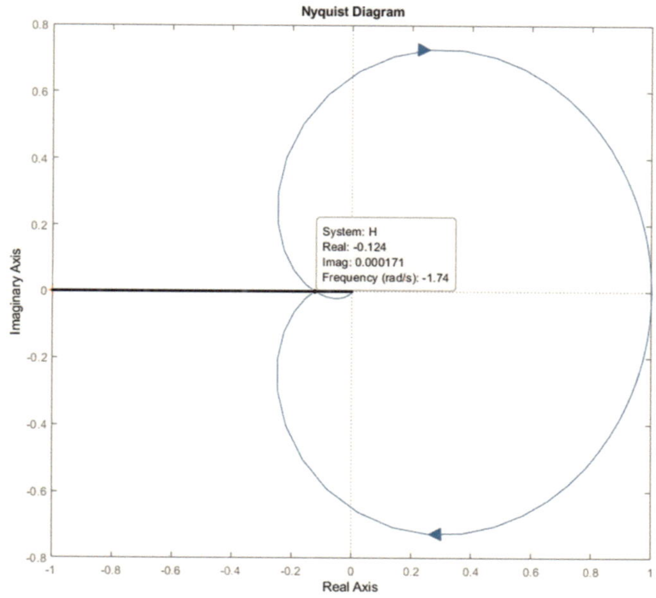

Figure 5.
The characteristics for harmonic analysis.

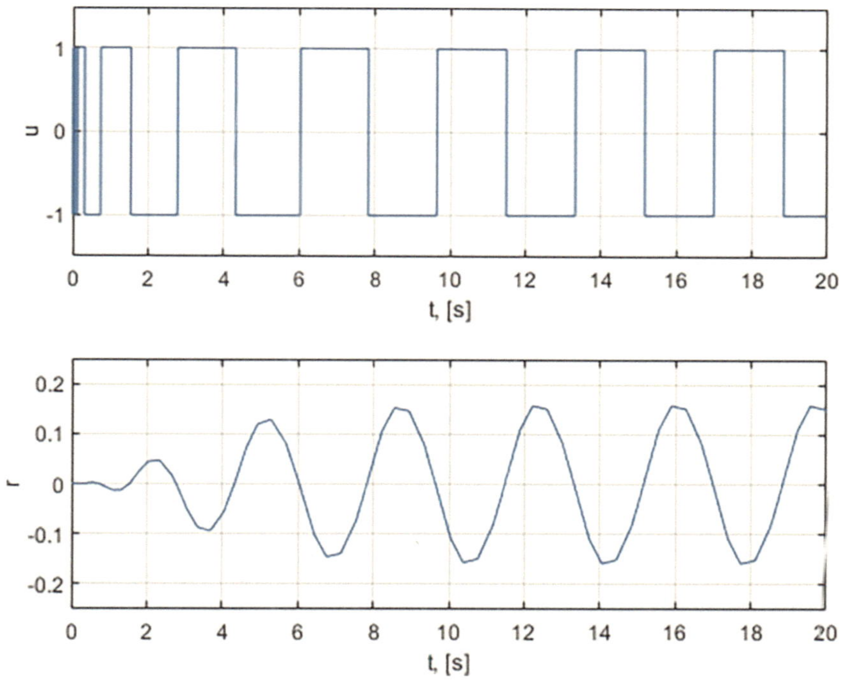

Figure 6.
Cycle oscillation representation for control system variables.

3. Nonlinear systems described with nonlinear differential equations

Often the nonlinear systems are described with nonlinear differential equations [10]

Figure 7.
Expressive examples for the stability of equilibrium points.

$$\dot{x} = f(x, u, t),\ x \in R^n, u \in R^m,\ t \in [0, \infty),\ x(0) = x_0 \tag{11}$$

To study the internal stability of a system, $u = 0$ is considered. The system is autonomous when t does not appear in f. The solutions of the form $x(t) = c$, where c is a constant, are the *equilibrium points* of the system [11]. Nonlinear systems may have empty, finite or infinite equilibrium sets. The concept of *stability* is related to equilibrium point stability. The most expressive example is presented in **Figure 7** – a parabolic bowl and a ball.

Figure 7 shows suggestively the phenomenon of equilibrium points. In the case of stable equilibrium points, after the system is removed from the equilibrium point by an external command, when the command disappears, the system returns to the stable equilibrium point. In the case of unstable equilibrium points, the system does not return. The role of control systems in engineering is to make the closed-loop system stable.

The phase (state) portrait [12] is the totality of the trajectories of a system in the state plane. Next, some examples of state portraits for second-order linear systems with the state Eq. (12) and the equation in the state plain (13)

$$\dot{x}_1 = f_1(x_1, x_2),\ \dot{x}_2 = f_2(x_1, x_2), \tag{12}$$

$$\frac{dx_1}{dx_2} = \frac{f_1(x_1, x_2)}{f_2(x_1, x_2)} \tag{13}$$

with the eigenvalues λ_1 and λ_2 and Eq. (14)

$$\dot{x}_1 = \lambda_1 x_1,\ \dot{x}_2 = \lambda_2 x_2 \tag{14}$$

The sign of λ_1 and λ_2 determines the stability and the shape of the phase portrait. Below are some critical points of this system, accompanied by their graphic representations in **Figure 8**: (a) stable node $\lambda_1, \lambda_2 < 0$; (b) unstable node $\lambda_1, \lambda_2 > 0$; (c) saddle point $\lambda_1 > 0,\ \lambda_2 < 0$; (d) stable focus $\lambda_{1,2} = a \pm bi \in C, a < 0$; (e) unstable focus $\lambda_{1,2} = a \pm bi \in C, a > 0$ and (f) center$\lambda_{1,2} = \pm bi \in C, a = 0$.

4. Stability

Second Lyapunov theorem [1, 2, 6]. Let

$$\dot{x} = f(x),\ f: D \to R^n \tag{15}$$

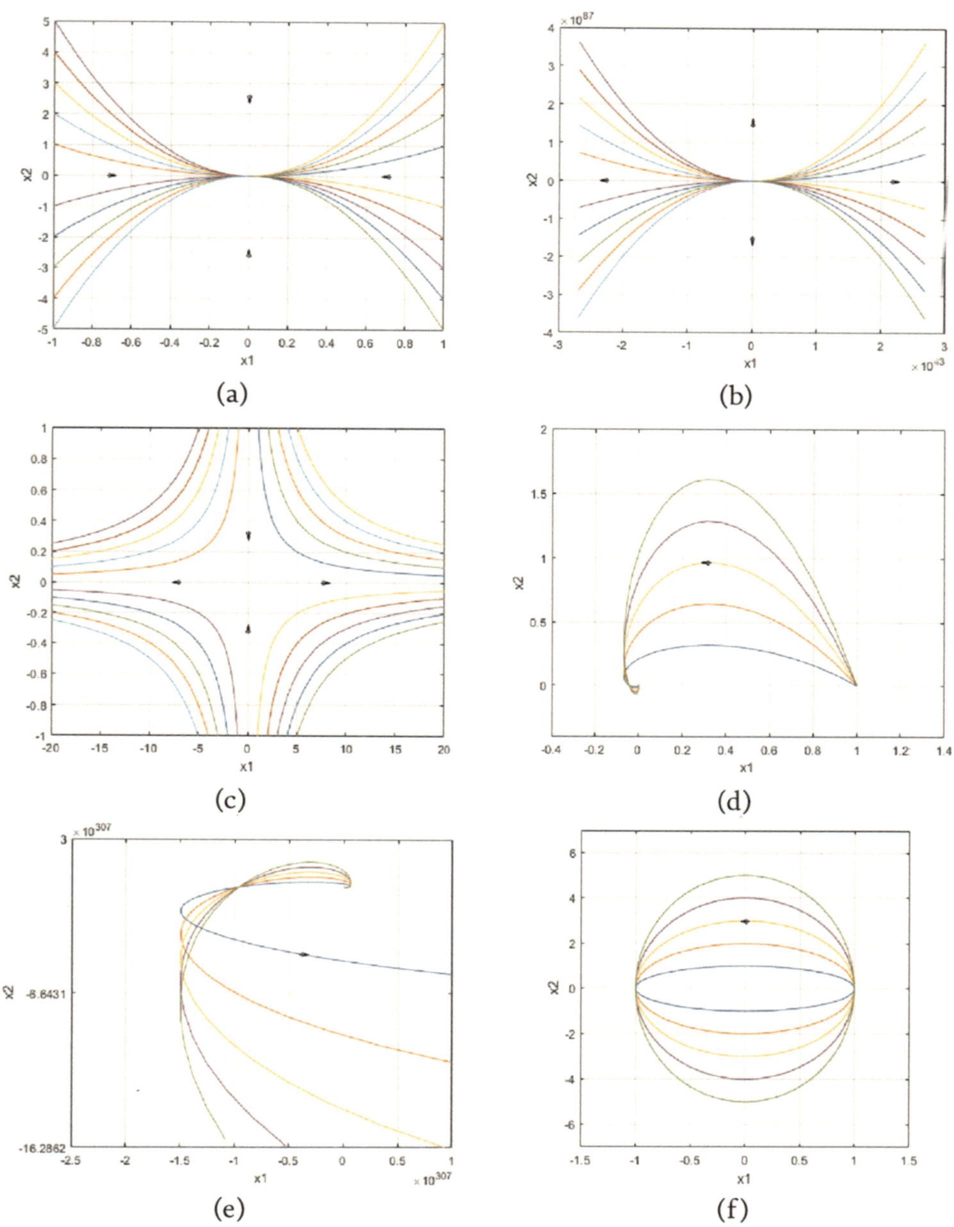

Figure 8.
State portrait graphs for critical points.

be a nonlinear system. Let $V(x) : D \to R^n$ be a continuously differentiable function positive definite $V(x) > 0,\ in\ D - \{0\}$ and $V(0) = 0$. If $\dot{V}(x) \leq 0$, the system is stable; $\dot{V}(x) < 0$, the system is asymptotically stable; $\lim_{\|x\| \to \infty} V(x) = \infty$, the system is global asymptotically stable. For a system with equations of state

$$\dot{x} = Ax \tag{16}$$

the derivative of the *Lyapunov function* is of the form

$$\dot{V}(x) = x^T\left[A^T P + PA\right]x \tag{17}$$

where P is a symmetric and positive definite matrix. A general recommendation is taking into account the physical phenomenon in the system and to choose V as the total energy of the system. If $\dot{V}(x) < 0$, it is negative semi-definite; this means the energy of the system is not increasing. If $\dot{V}(x) \leq 0$, it is negative, implies that the energy is decreasing and the system is asymptotically stable.

Example. The theorem is applied to analyze the behavior of the pendulum. The schematic description that characterized the motion of the pendulum is given in **Figure 9**, where: m is the mass of the pendulum, R is the arm length, $G = mg$ is the gravitational force, v is the tangential velocity, ω is the rotational velocity, θ is the angle with the vertical, F_c is the centripetal force, F_1 and F_2 are the components of G along the two perpendicular axes.

The system state variables and the state equations are

$$x_1 = \theta,\ x_2 = \omega \tag{18}$$

$$\dot{x}_1 = -x_2,\ \dot{x}_2 = \frac{g}{R}\sin x_1 \tag{19}$$

with the equilibrium point

$$x_{1e} = \theta_e = 0,\ x_{2e} = \omega_e = 0 \tag{20}$$

The recommended Lyapunov function is the total energy of the system, which is the sum of the kinetic energy and the potential energy

$$V = E_t = E_c + E_p = \frac{mR^2}{2}\omega^2 + mgh = \frac{mR^2}{2}x_2^2 + mgR(1 - \cos x_1) > 0 \tag{21}$$

Total energy of a system is positive.

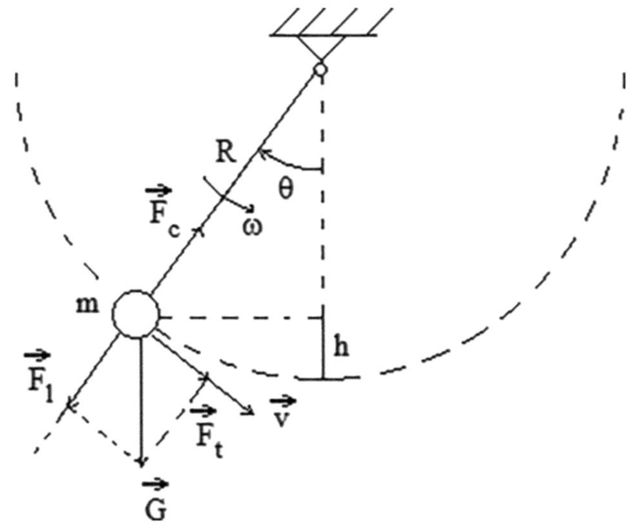

Figure 9.
Pendulum motion diagram.

The derivative of V is the derivative of a multivariable function

$$\dot{V} = \frac{\partial V}{\partial x_1}\dot{x}_1 + \frac{\partial V}{\partial x_2}\dot{x}_2 = mgR(-x_2)\sin x_1 + mR^2 x_2 \frac{g}{R}\sin x_1 = 0 \tag{22}$$

The system is stable because there is a small amount of friction that causes the energy to decrease and the system to reach equilibrium. The equilibrium point is stable. The system has in practice a small friction, which causes the derivative of V to be negative and its motion to be damped.

The equilibrium point, which has coordinates in the plane of states [θ, ω], should not be confused with the geometric point of the pendulum's position in the plane.

The equilibrium point of the inverted pendulum (θ = 180°, ω = 0) is unstable.

Phase portrait. The equation of the pendulum in state space is further deduced

$$\frac{dx_1}{dx_2} = -\frac{R}{mg}\cdot\frac{x_2}{\sin x_1} \Rightarrow \sin x_1 dx_1 = -\frac{R}{mg}x_2 dx_2 \Rightarrow -\cos x_1 = -\frac{R}{2mg}x_2^2 \tag{23}$$

The solution is

$$x_1 \in \left[-\frac{\pi}{2}, \frac{\pi}{2}\right],\ x_2 = \pm\sqrt{\frac{2mg}{R}\cos x_1} \tag{24}$$

The state portrait is presented in **Figure 10**. The equilibrium point (31) is a center.

Popov's criterion [13] is a graph-analytic method for studying the stability of nonlinear systems. The process P in Popov's criterion may have the transfer function

$$H(s) = \frac{1}{s^p}\cdot\frac{b_m s^m + b_{m-1}s^{m-1} + \ldots + b_0}{a_n s^n + a_{n-1}s^{n-1} + \ldots + a_0} \tag{25}$$

The nonlinearity N should have the characteristic

$$u = N(e), 0 < \frac{N(e)}{e} < K, N(0) = 0 \tag{26}$$

and its graph must fit into the Popov sector as in **Figure 11**.

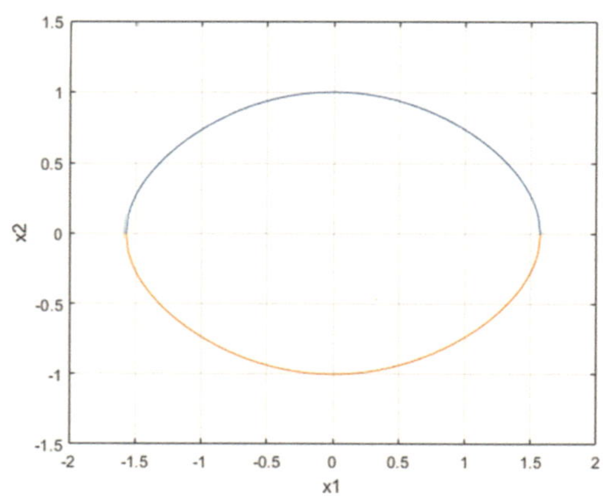

Figure 10.
Pendulum state portrait.

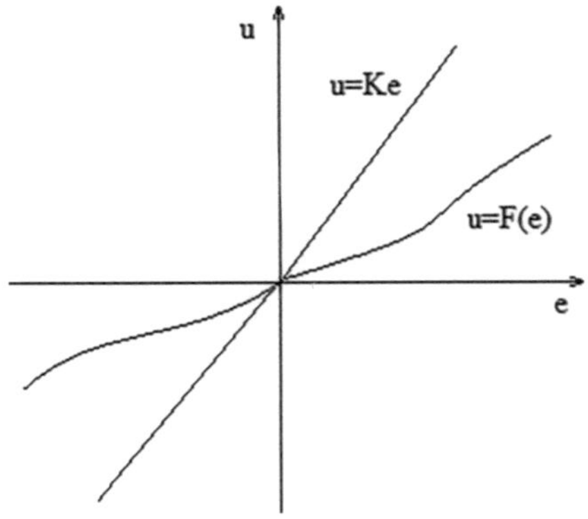

Figure 11.
Nonlinearity graph.

According to Popov's criterion, the closed loop is absolutely stable in the sector $[0, K]$, if there is a real number q, so for any $\omega \geq 0$, to satisfy the Popov's inequality

$$\text{Re}\{(1 + j\omega q)H(j\omega)\} + \frac{1}{K} > 0 \tag{27}$$

Popov's hodograph has the equation

$$H_P(j\omega) = \text{Re}\{H(j\omega)\} + j\omega\text{Im}\{H(j\omega)\} = a + jb \tag{28}$$

The condition in the theorem translates into Popov's inequality

$$a - qb + 1/K > 0 \tag{29}$$

This condition is obtained for all the points at the right of the Popov's straight line

$$a - qb + 1/K = 0 \tag{30}$$

The slope of the Popov's line is $1/q$. Punctual intersection with the real axis is $-1/K$.

So, the Popov's hodograph must be situated at the right of the critical Popov's straight line. The maximum Popov sector $[0, K_P]$ may be included in the Hurwitz sector $[0, K_H]$.

Example. Let a linear system with transfer function

$$H(s) = \frac{1}{s} \cdot \frac{1}{0.001s^2 + 0.01s + 1} \tag{31}$$

The hodograph and Popov's line are presented in **Figure 12**.

The intersection with x axis has the abscissa $-1/K = -0.0909$. The maximum value of Popov sector limit is $K_P = 10.989$. So, the closed-loop control is absolutely stable in the sector $[0, K_P]$, if there is a real number q, to satisfy the Popov's inequality (27).

The upper edge of the Popov sector indicates the maximum value that the regulator amplification can take so that the control system remains stable: $K < K_P$. A change in process parameters modifies Popov's hodograph over time and leads to instability. To ensure stability, it is recommended to ensure a sufficient tolerance margin for K. The nonlinearities (b) and (c) from Section 2 can fit into a Popov sector because they

DOI: http://dx.doi.org/10.5772/intechopen.1013515

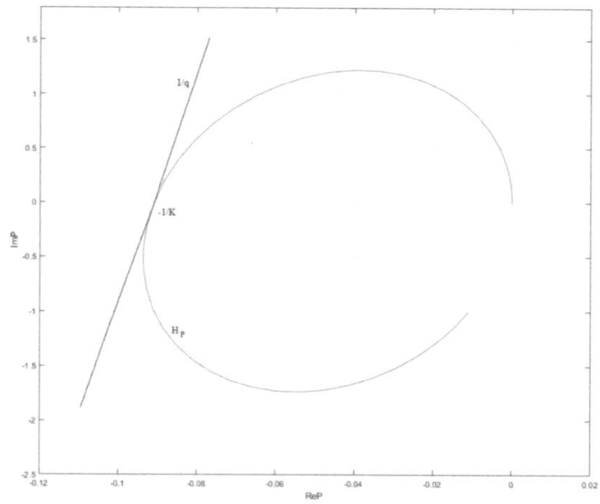

Figure 12.
Graphic representation of Popov's criterion application.

fulfill the Popov sector condition. Nonlinearities (a) and (d) cannot fit into a Popov sector, as in **Figure 11**, and they produce limit cycles in control systems because $N(0) \neq 0$.

5. Conclusion

The chapter presents some notions about essential nonlinearities, accompanied by an example of modeling and simulation of a control system of temperature, with a nonlinear regulator, often used in practice. The occurrence of limit cycles is highlighted. For the analysis of limit cycles, the following notions are presented: the harmonic balance method, the definition of the descriptive function for common static nonlinearities, the harmonic balance equation used in practice and the calculation procedure with some examples with modeling and simulation to demonstrate the theory. The harmonic balance method is an approximate method, but very easy to use. Introductory notions for systems with nonlinearities described with differential equations are presented on a discussion about equilibrium points. Notions about the state portrait of systems are presented with graphic examples for the decoupled second-order linear system and for the pendulum. Notions regarding Lyapunov's second stability theorem are presented on an example of pendulum stability analysis. Lyapunov theorem can be applied without solving state equations. Notions regarding Popov's criterion are presented with an example. Popov's criterion is an approximate graphic method, but very easy to use. MATLAB was used for modeling and simulation.

Conflict of interest

The author declares no conflict of interest.

Author details

Constantin Volosencu
Politehnica University, Timisoara, Romania

*Address all correspondence to: constantin.volosencu@aut.upt.ro

References

[1] Khalil H. Nonlinear Systems. 3rd ed. New York City NY: Pearson; 2001. 768 p

[2] Isidori A. Nonlinear Control Systems. 3rd ed. Berlin: Springer; 1995. 549 p. ISBN 978-3-540-19916-8

[3] Volosencu C. Stability analysis of systems with fuzzy PI controllers applied to electric drives. Mathematics. 2021; **19**(9):1246. DOI: 10.3390/math9111246

[4] Volosencu C. Study of the angular positioning of a rotating object based on some computational intelligence methods. Mathematics. 2022;**10**(7):1157. DOI: 10.3390/math10071157

[5] Volosencu C. A comparative analysis of some methods for wind turbine maximum power point tracking. Mathematics. 2021;**9**(19):2399. DOI: 10.3390/math9192399

[6] Yoshizawa T. Stability Theory by Liapunov's Second Method. Vol. 9. Tokyo: Mathematical Society of Japan; 1966. 223 p

[7] Dutta D, Bhattacharjee JK. Limit Cycle Oscillations. In: İnan E, Sengupta D, Banerjee M, Mukhopadhyay B, Demiray H, editors. Vibration Problems ICOVP-2007. Springer Proceedings in Physics. Vol. 126. Dordrecht: Springer; 2008. pp. 125-140. DOI: 10.1007/978-1-4020-9100-1_14

[8] Jazar RN. Harmonic balance method. In: Jazar RN, editor. Perturbation Methods in Science and Engineering. Dordrecht: Springer; 2021. pp. 263-353. DOI: 10.1007/978-3-030-73462-6_4

[9] Usai E. Describing Function: Analysis of Nonlinear Systems [Internet]. Rode, India: Erode Sengunthar Engineering College; 2024. p. 84. Available from: https://www.electrical4u.com/different-types-nonlinearities-in-control-system/ [Accessed: August 14, 2025]

[10] Hirsch MW, Smale S, Devaney RL. Differential Equations, Dynamical Systems, and an Introduction to Chaos. 3rd ed. Amsterdam, Nederland: Elsevier; 2013. 418 p. DOI:10.1016/C2009-0-61160-0

[11] Mostoufi N, Constantinides A. Applied Numerical Methods for Chemical Engineers. Cambridge, Mass: Academic Press; 2023. 492 p. DOI: 10.1016/C2019-0-04646-8

[12] Mellodge P. A Practical Approach to Dynamical Systems for Engineers. London: Elsevier; 2016. pp. 147-213. DOI: 10.1016/B978-0-08-100202-5.00003-6

[13] Popov VM. Hiperstabilitatea sistemelor automate. Bucuresti: Ed. Academiei; 1966. p. 303

Chapter 2

Sensor Fault Detection and Compensation Scheme for Vector-Controlled Induction Motor Drives

Fatiha Zidani

Abstract

This chapter presents a sensor fault-tolerant control (FTC) strategy for induction motor (IM) drives using an indirect field-oriented control (IFOC) approach based on rotor flux orientation. The proposed scheme ensures continued speed control in the presence of current sensor faults by employing a multi-observer architecture derived from Kubota observers. These observers provide reliable flux estimation even when partial current measurements are unavailable. A fault detection, isolation, and compensation process are integrated into the control scheme to maintain system stability and performance. Experimental results under fault scenarios demonstrate the effectiveness and robustness of the approach. The simplicity of the structure and its modular integration make it suitable for industrial applications requiring high reliability.

Keywords: induction motor, fault detection and isolation, fault-tolerant control, vector control, Kubota observer, rotor flux estimation, reliability

1. Introduction

Induction motors (IMs) are extensively utilized across various industrial sectors due to their robust nature, high efficiency, and cost-effectiveness. Nevertheless, ensuring the reliability of drive systems remains a significant challenge, particularly in safety-critical applications like transportation, robotics, and renewable energy. Sensor malfunctions, especially those affecting current sensors, can profoundly impact the performance and stability of field-oriented control (FOC) schemes.

Field-oriented control is a widely adopted technique for achieving high-performance control of IMs. It effectively decouples torque and flux dynamics by transforming stator currents into a rotating reference frame. Among the various vector control approaches, this chapter specifically examines the Indirect Field-Oriented Control (IFOC) method, which relies on rotor flux orientation. IFOC is favored for its straightforward implementation and has proven practical reliability.

Achieving fault tolerance in such systems necessitates three key capabilities: fault detection, fault isolation, and fault compensation. Conventional methods often struggle to maintain satisfactory control performance when one or more sensors fail. To address this, observer-based techniques, such as Kubota observers, can estimate rotor flux using only partial current measurements. This capability allows for continued operation even in the presence of sensor faults [1–8].

This work introduces a fault-tolerant control (FTC) architecture that integrates multiple observers with a vector control algorithm and a fault detection mechanism. The proposed system is designed to automatically detect faulty current sensors, isolate them through residual analysis, and then compensate for the fault by leveraging remaining healthy sensors and observer-based estimations. Both the theoretical framework and experimental validation under fault conditions are presented in this chapter.

The following sections are organized as follows:

Section 2 presents the dynamic model of the induction machine.

Section 3 presents the IFOC principles.

Section 4 introduces the fault-tolerant control (FTC) architecture.

Section 5 details the flux observers, fault detection, and isolation algorithm.

Section 6 describes the hardware setup and implementation.

Section 7 presents and discusses the experimental results.

Section 8 concludes the chapter and highlights future directions.

2. Mathematical model of the induction motor

In the d-q synchronous reference frame, the state-space equations of the induction motor are expressed as follows:

$$\begin{cases} V_{sd} = R_s I_{sd} + \dfrac{d\phi_{sd}}{dt} - \omega\phi_{sq} \\ V_{sq} = R_s I_{sq} + \dfrac{d\phi_{sq}}{dt} + \omega\phi_{sq} \\ 0 = R_r I_{rd} + \dfrac{d\phi_{rd}}{dt} \\ 0 = R_r I_{rq} + \dfrac{d\phi_{rq}}{dt} \end{cases} \quad (1)$$

The flux linkage equations are:

$$\begin{cases} \phi_{sd} = L_d I_{sd} + M_d I_{rd} \\ \phi_{qs} = L_q I_{sq} + M_q I_{rq} \\ \phi_{rd} = L_d I_{rd} + M_d I_{sd} \\ \phi_{rq} = L_d I_{rq} + M_d I_{sq} \end{cases} \quad (2)$$

Where V_{sd} and V_{sq} are the d- and q-axis voltages, R_s is the stator resistance, i_{sd} is the direct current, i_{sq} is the quadratic current, ϕ_{sd} and ϕ_{sq} are the d- and q-axis flux linkages, ω is the rotor speed. L_d is the d-axis inductance, L_q is the q-axis inductance, M_d is the d-axis mutual inductance, and M_q is the q-axis mutual inductance (**Figure 1**).

By handling Eqs. (1) and (2), we obtain the following flux model:

$$\begin{cases} \dfrac{dI_{sd}}{dt} = \dfrac{-R_s}{L_d} I_{sd} + \omega\dfrac{L_q}{L_d} I_{sq} + \dfrac{1}{L_d} V_{sd} \\ \dfrac{dI_{sq}}{dt} = \dfrac{-R_s}{L_q} I_{sq} - \omega\dfrac{L_d}{L_q} I_{sd} + \dfrac{1}{L_q} V_{sq} \end{cases} \quad (3)$$

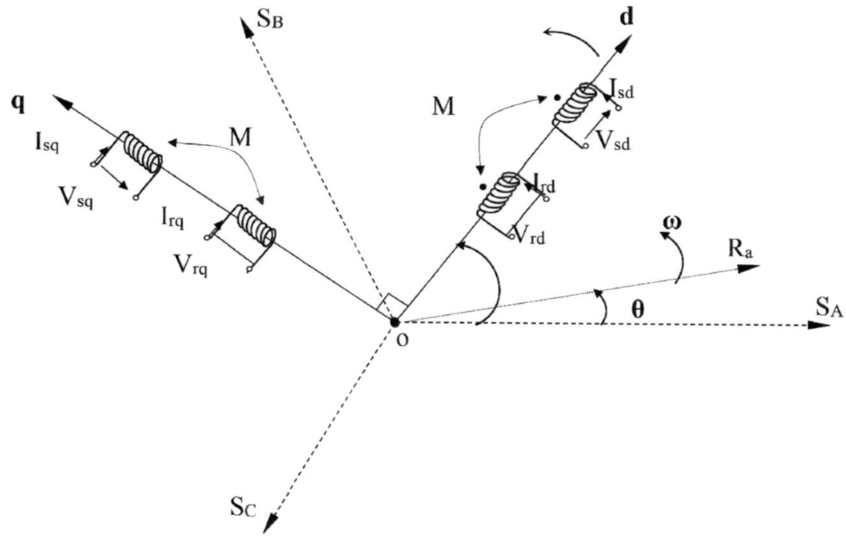

Figure 1.
Winding representation of IM in (d, q) reference.

The electromagnetic torque can be expressed as

$$\Phi_{rd} = \frac{3p}{2}\left(\phi_{sd}I_{sq} - \phi_{sq}I_{sd}\right) \quad (4)$$

Where Γ_e is the electromagnetic torque of the motor, and p is the number of poles of the induction motor. The rotor speed and position of the motor can be expressed as:

$$\frac{d\omega_r}{dt} = \frac{1}{J}(\Gamma_e - T_l - f\omega_r) \quad (5)$$

and

$$\frac{d\theta}{\mathrm{dt}} = \omega_r \quad (6)$$

where J is the inertia constant of the motor and load, Γ_L is the external load torque, f is the viscous frictional coefficient of the motor and load, θ is the mechanical rotor position, and ω is the mechanical rotor speed.

3. Indirect field-oriented control (IFOC) principles

In conventional AC motor control, a significant challenge arises from the inherent coupling between the variables responsible for producing magnetic flux and mechanical torque. This means that any attempt to alter one variable, for instance, by adjusting voltage or frequency, inadvertently impacts the other. Such interdependence leads to highly nonlinear and sluggish responses, particularly during transient operating conditions. This coupling fundamentally complicates the achievement of precise and dynamic control over the motor's behavior. The complex mathematical model and non-linear characteristics of induction motors further exacerbate these control difficulties [9].

This fundamental coupling represents a significant barrier to achieving high-performance AC drives. Despite their inherent robustness, AC motors traditionally suffered from poor dynamic performance. The root cause of this limitation is explicitly identified as the strong coupling between flux and torque variables. This implies that a simple adjustment of input parameters, as in scalar control, does not isolate the effect on torque; it simultaneously alters the magnetic field, making it exceedingly difficult to achieve precise control. Consequently, this coupling results in non-linear behavior, slow response times, and undesirable torque oscillations, which historically confined AC motors to fixed-speed or less demanding applications. Understanding this fundamental control problem is essential for appreciating why FOC emerged as a necessary and revolutionary development, transforming AC motors from basic workhorses into precision instruments capable of dynamic and accurate operation.

DC motors have historically been favored for high-performance applications due to a distinct advantage: their magnetic field, controlled by the field current, and their armature current, which produces torque, are naturally orthogonal and thus inherently decoupled. This intrinsic separation allows for independent control of both flux and torque, significantly simplifying the control scheme and enabling a highly precise and dynamic response. In a DC motor, the field flux linkage and armature magnetomotive force (m.m.f) maintain an orthogonal alignment, irrespective of speed [9].

The primary objective of FOC is to replicate this ideal DC motor behavior within AC machines. By achieving this "DC motor-like" performance, FOC transforms the complex, non-linear AC motor into a linear control plant when viewed within the transformed reference frame. This linearization is a profound simplification, as it then permits the use of simpler, well-understood linear control techniques, such as Proportional-Integral (PI) controllers, for the independent regulation of flux and torque. This conceptual framework, where the AC motor is mathematically engineered to behave like a DC motor, not only clarifies the design of intricate control systems but also renders the motor's behavior far more predictable and manageable, making it a powerful tool for advanced applications.

3.1 Decoupled torque and flux control

There are two primary methods for implementing vector control: Direct Field-Oriented Control *DFOC* and Indirect Field-Oriented Control *IFOC*. These approaches differ fundamentally in how they determine the rotor flux angle.

Direct Field-Oriented Control *DFOC*: This method directly estimates the rotor flux angle, often utilizing physical flux sensors, such as Hall sensors or search coils, or by employing voltage/current models. While *DFOC* is conceptually straightforward in its reliance on direct measurement, the use of physical flux sensors can introduce several limitations. These sensors may be mechanically unstable, susceptible to noise interference, and typically offer only localized sensing of the flux, which may not accurately represent the total rotor flux.

Indirect Field-Oriented Control *IFOC*: More commonly adopted in closed-loop drive systems, *IFOC* estimates the rotor angle from the slip frequency and the rotor frequency, typically without the need for direct flux sensors. In this approach, the slip frequency is calculated based on the motor's parameters and measured stator currents. This calculated slip frequency is then summed with the rotor angle (obtained from a speed sensor or an estimator) to derive the flux space angle. *IFOC* generally offers superior dynamic performance, particularly at low speeds and frequencies.

The prevalent preference for *IFOC* in many applications is a direct consequence of its ability to eliminate expensive and potentially fragile sensors, despite the heightened reliance on accurate parameter estimation and the subsequent need for adaptive algorithms to compensate for parameter drift. This choice between *DFOC* and *IFOC* reflects a common engineering dilemma: balancing performance, cost, and robustness. The increasing trend toward sensorless *IFOC* underscores the industry's continuous drive for more robust, simpler, and cost-effective solutions, pushing the boundaries of real-time estimation techniques.

This task *IFOC* is achieved by transforming the stator currents into a rotating reference frame (d-q axes) that is aligned with the chosen flux. In this frame, the d-axis current i_{sd} controls the flux, and the q-axis current i_{sq} controls the torque, similar to a *DC* motor. This simplification allows for independent control and improved dynamic performance.

Regarding the "three possible choices of flux orientation" in vector control, these refer to the different magnetic flux linkages that can be chosen as the reference for aligning the d-axis. These choices are:

a. Orientation du Flux rotorique: $\phi_{rd} = \phi_r$ et $\phi_{rq} = 0$

b. Orientation du Flux statorique: $\phi_{sd} = \phi_s$ et $\phi_{sq} = 0$

c. Orientation du Flux d'entrefer:: $\phi_{ed} = \phi_e$ et:$\phi_{ed} = \phi_e$

Rotor flux orientation: This is the most common and widely used method in IFOC. In this scheme, the d-axis of the rotating reference frame is aligned with the rotor flux vector. This choice simplifies the control equations and provides good dynamic performance, especially for induction motors where the rotor flux is crucial for torque production.

Stator flux orientation: In this method, the d-axis is aligned with the stator flux vector. While it can also provide decoupled control, it is less common for general-purpose induction motor drives compared to rotor flux orientation.

Magnetizing flux orientation (or air-gap flux orientation): Here, the d-axis is aligned with the magnetizing flux (or air-gap flux) vector. This approach can be used, but it is generally more complex to implement than rotor flux orientation.

The choice of flux orientation impacts the complexity of the control algorithm and the performance characteristics of the drive system. For IFOC, rotor flux orientation is typically preferred due to its practical advantages [9]. Rotor flux orientation is the method of choice, as shown in the following **Figure 2**.

From the rotor flux orientation properties, we obtain the following relationships:

$$\begin{cases} \phi_{rq} = 0 \\ \phi_{rd} = \phi_r = cst \end{cases} \tag{7}$$

we obtain the following current equations:

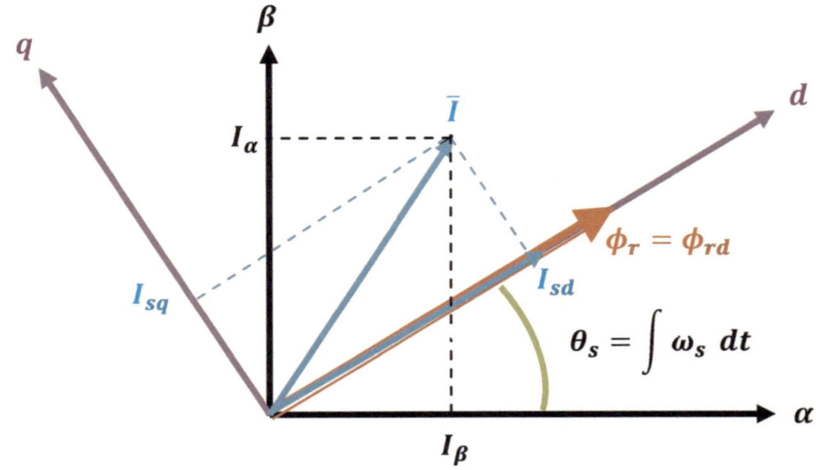

Figure 2.
Rotor flux orientation.

$$I_{rd} = \frac{1}{L_r}(\phi_r - MI_{sd}) \tag{8}$$

$$I_{rq} = \frac{-M}{L_r} I_{sq} \tag{9}$$

From equations the above equations we obtain

$$V_{sd} = \sigma L_s \frac{dI_{sd}}{dt} + \left(R_s + R_r \frac{M^2}{L_r{}^2}\right) I_{sd} - \omega_s \sigma L_s I_{sq} - \frac{M}{L_r{}^2} R_r \phi_r \tag{10}$$

$$V_{sq} = \sigma L_s \frac{dI_{sq}}{dt} + \left(R_s + R_r \frac{M^2}{L_r{}^2}\right) I_{sq} + \omega_s \sigma L_s I_{sd} + \omega_r \frac{M}{L_r} \phi_r \tag{11}$$

3.2 Compensation decoupling

We define two new control variables V_{sd} and V_{sq}. V_{sd} acts only on I_{sd} and V_{sq} acts on I_{sq}. V_{sd} et V_{sq} are defined as follows:

$$\begin{cases} V_{sd} = V_{sd} - e_q \\ V_{sq} = V_{sq} - e_d \end{cases}$$

with
$e_d = \omega_s \sigma L_s I_{sq} + \frac{M}{L_r{}^2} R_r \phi_r$, $e_q = -\omega_s \sigma L_s I_{sd} - \frac{M}{L_r} \omega_r \phi_r.$
and
$V_{sd} = \sigma L_S \frac{dI_{sd}}{dt} + \gamma I_{sd}$, $V_{sq} = \sigma L_S \frac{dI_{sq}}{dt} + \gamma I_{sq}$
with $\gamma = \left(R_s + R_r \frac{M^2}{L_r{}^2}\right)$

The voltages V_{sd} and V_{sq} are then reconstituted from V_{sd} and V_{sq} to achieve perfect decoupling between the (d, q) axes.

e_d et e_q: disturbance e.m.f. related to currents and introducing nonlinear coupling.

The machine model can be summarized as follows:

$$\begin{cases} V_{sd} = (R_s + P\sigma L_s)I_{sd} - \underbrace{\sigma L_s \omega_s I_{sq}}_{f.e.m - \mathrm{q}} + \dfrac{M}{T_r} P\varphi_r \\ V_{sq} = (R_s + P\sigma L_s)I_{sq} + \underbrace{\sigma L_s \omega_s I_{sd} + \dfrac{M}{L_r}\omega_s \varphi_r}_{f.e.m - \mathrm{d}} \\ \dfrac{\varphi_r}{I_{\mathrm{sd}}} = \dfrac{M}{1 + T_r P} \\ \Gamma_e = n_p \dfrac{M}{L_r}\varphi_r I_{sq} \\ \Omega(JP + f) = T_e - T_L \\ P\theta_s = \omega_s = \omega + \dfrac{M I_{sd}}{T_r \varphi_r} \end{cases} \tag{12}$$

It can be noted that only the direct-axis "d" axis component of the stator current determines the amplitude of the rotor flux, whereas the torque depends on the quadrature "q" axis component of the stator current. If the flux is maintained constant, the resulting control structure becomes similar to that of a naturally decoupled DC machine.

3.3 Control estimators

The direct-axis stator current I_{sd} controls the amplitude of the rotor flux Φ_{rd}.

The quadrature-axis stator current I_{sq} determines the developed torque Γ_e.

If the flux is kept constant, the control strategy of an AC machine using *IFOC* becomes analogous to that of a *DC* machine, where torque is directly proportional to armature current and flux is separately controlled.

For the ideal decoupling state, the developed torque is defined as follows:

$$\Gamma_e = n_p \frac{M}{L_r} \Phi_{rd} i_{sq} \tag{13}$$

Where Φ_{rd} can be estimated by the measured stator current:

$$\hat{\Phi}_{rd} = \frac{M}{1 + T_r . p} i_{sdref} \tag{14}$$

The rotor flux position θ_s required for the coordinates transformation is calculated from the rotor speed $n_p \omega$ and slip frequency $\hat{\omega}_r$.

$$\theta_s = \int \omega_s \mathrm{dt} \text{ where } \omega_s = n_p \omega + \hat{\omega}_r \tag{15}$$

$\hat{\omega}_r$ is derived from the stator current, the estimated rotor and the motor parameters:

$$\hat{\omega}_r = \frac{M}{T_r . \hat{\Phi}_{rd}} i_{sqref} \tag{16}$$

A conventional IFOC system typically employs two cascaded PI (Proportional-Integral) controllers:

One for speed regulation: This controller generates the torque command (and thus the q-axis current reference).

One for current control: These controllers regulate the d-axis and q-axis currents by manipulating the stator voltage commands.

The output voltage vector is generated using Space Vector PWM (Pulse Width Modulation) techniques. **Figure 3** illustrates this IFOC scheme with classical PI controllers.

4. FOC fault-tolerant control strategy *FTC* architecture

A system's ability to continue functioning despite failures, a concept known as fault tolerance, has become a key requirement in many engineering domains. The primary objective of Fault-Tolerant Control (FTC) is to ensure system resilience by guaranteeing its stability and limiting performance degradation when facing malfunctions. Traditional control strategies often show their limitations in such situations. Therefore, new control architectures, integrating fault detection and compensation mechanisms, have been developed. The integration of state observers, like the Kubota observer for rotor flux, is fundamental for implementing these advanced strategies, particularly in field-oriented control systems.

The Kubota observer is a popular method to observe non measurable variables (rotor flux) [1, 2]. **Figure 4** shows the proposed architecture for fault control system [10–12]. The proposed scheme incorporates FOC algorithm and rotor flux observers. Observed flux and fault detection with the selected variables are feedback on the control system based on IFOC method.

4.1 Overview of the fault-tolerant control (FTC) strategy

To ensure uninterrupted operation of induction motors despite sensor failures, a robust Fault-Tolerant Control (FTC) strategy is indispensable. The proposed method is characterized by:

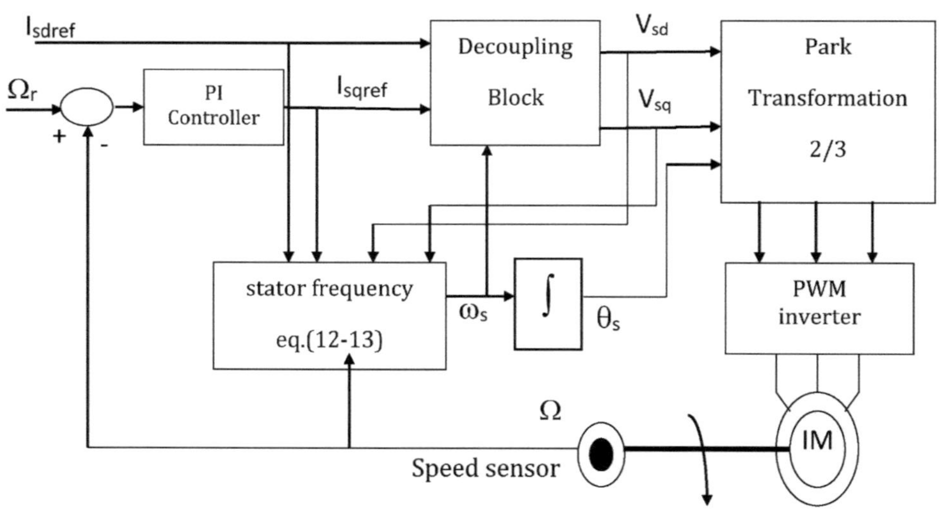

Figure 3.
Block diagram of indirect field-oriented induction motor drive.

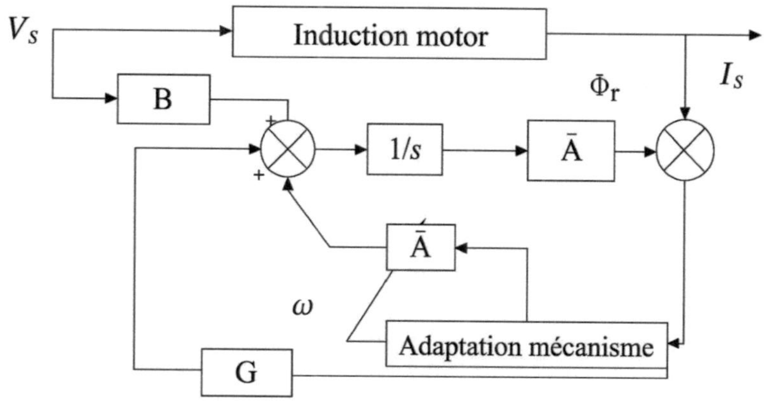

Figure 4.
Block diagram of the KUBOTA adaptive observer.

The use of multiple Kubota observers, offering redundancy in motor flux estimation and enabling compensation for lost current measurements.

A fault detection and isolation (FDI) module that generates significant residuals indicating sensor status.

An intelligent compensation mechanism capable of switching to an operational observer configuration in the presence of a faulty sensor. This architecture is directly embedded within the IFOC algorithm (see **Figure 5**), allowing for real-time monitoring of sensor integrity and autonomous reconfiguration to preserve control quality. Specifically, the system relies on three Kubota flux observers, each leveraging different current combinations, decision logic for identifying discrepancies, residual generators for evaluating sensor health, and a switching block for dynamic selection.

The different tasks of the FTC block are:

1. Redundant flux observers (Kubota observers) to estimate motor states even with missing current measurements.

2. A fault detection and isolation (FDI) mechanism that generates residual signals.

3. A compensation mechanism that switches to an alternative observer configuration when a sensor is faulty.

The system monitors sensor health in real time and reconfigures itself automatically to maintain performance. The proposed architecture includes:

- Three Kubota flux observers, each using a different combination of two-phase currents. A decision logic unit to compare outputs and detect inconsistencies.

- Residual generators to evaluate sensor integrity.

- A switching block that selects a healthy observer configuration.

4.2 The KUBOTA adaptive observer

As illustrated in **Figure 4**, the adaptive KUBOTA observer is a crucial component of this strategy. Its operation is based on the exploitation of the induction machine's vector state model, formulated in the stator reference frame. Within this framework, the rotor flux vector and the stator current vector are the fundamental state variables that the observer strives to estimate accurately

4.2.1 IM model in the (α, β) reference frame linked to the stator

We will now proceed to the state equations of the machine model that will be used to design our observer. Thus, the IM model is described by the following equation of state:

$$\begin{cases} \dot{X} = A(\omega)X + BU \\ Y = CX \end{cases} \tag{17}$$

such as:

$$X = \left[I_{s\alpha} I_{s\beta} \phi_{r\alpha} \phi_{r\beta}\right]^T, Y = I_s \left[I_{s\alpha} I_{s\beta}\right]^T \text{ et } U = V_s = \left[V_{s\alpha} V_{s\beta}\right]^T$$

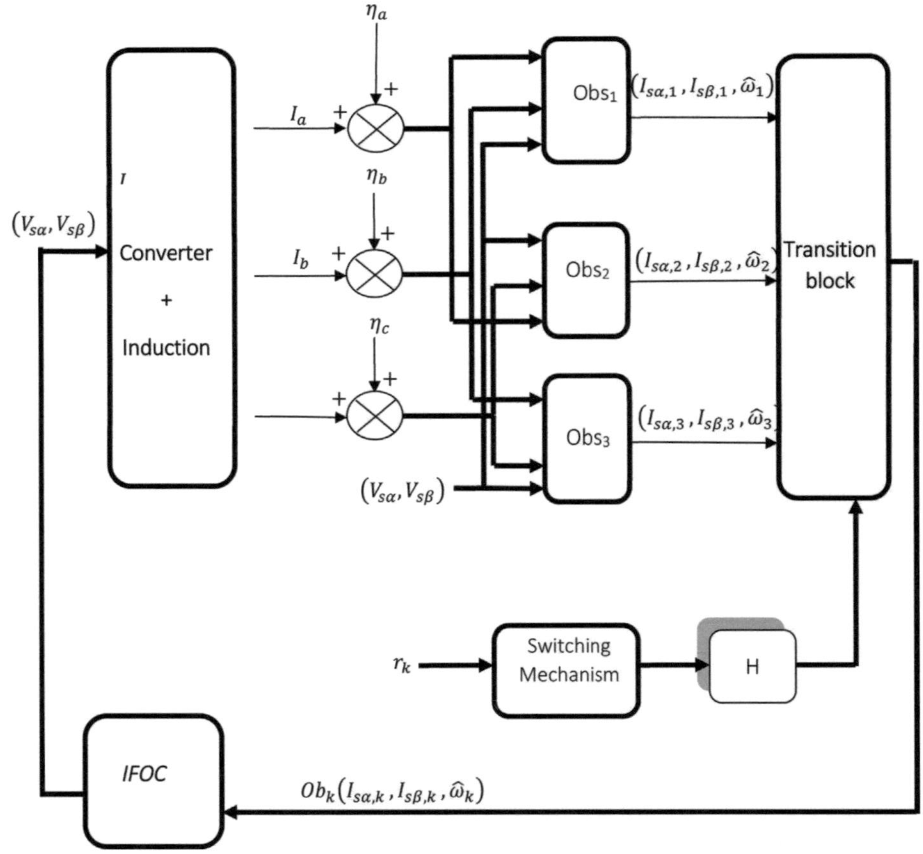

Figure 5.
Scheme of fault-tolerant control system.

with:

$$[A] = \begin{bmatrix} -\alpha_1 & 0 & \alpha_2 & p\alpha_2 T_r\omega \\ 0 & -\alpha_1 & p\alpha_2 T_r\omega & \alpha_2 \\ \alpha_4 & 0 & -\alpha_5 & p\omega \\ 0 & \alpha_4 & p\omega & -\alpha_5 \end{bmatrix}; [B] = \begin{bmatrix} \frac{1}{\sigma L_s} & 0 \\ 0 & \frac{1}{\sigma L_s} \\ 0 & 0 \\ 0 & 0 \end{bmatrix} \text{et } [C] = \begin{bmatrix} 1 & 0 & 0 & 0 \\ 0 & 1 & 0 & 0 \end{bmatrix}$$

with:

$$\alpha_1 = \left(\frac{1}{\sigma T_s} + \frac{(1-\sigma)}{\sigma T_r}\right), \alpha_2 = \frac{M}{\sigma L_s L_r T_r}, \alpha_3 = \frac{M}{\sigma L_s L_r}, \alpha_4 = \frac{M}{T_r}, \alpha_5 = \frac{1}{T_r}$$

4.2.2 State representation of KUBOTA observer

The purpose of an observer is to carry out a state feedback control and to estimate this state by a variable that we will call $\hat{X}$.

such as: $\hat{X} = \left[I_{s\alpha} I_{s\beta} \phi_{r\alpha} \phi_{r\beta}\right]^T$

$$\begin{cases} \dot{\hat{I}}_{s\alpha} = -\alpha_1 \hat{I}_{s\alpha} + \alpha_2 \hat{\phi}_{r\alpha} + p\alpha_2\omega\hat{\phi}_{r\beta} + \left(\frac{1}{\sigma L_s}\right) V_{s\alpha} + g_1\left(\hat{I}_{s\alpha} - I_{s\alpha}\right) - g_2\left(\hat{I}_{s\beta} - I_{s\beta}\right) \\ \dot{\hat{I}}_{s\beta} = -\alpha_1 \hat{I}_{s\beta} + \alpha_2 \hat{\phi}_{r\beta} - p\alpha_2\omega\hat{\phi}_{r\alpha} + \left(\frac{1}{\sigma L_s}\right) V_{s\alpha} + g_2\left(\hat{I}_{s\alpha} - I_{s\alpha}\right) + g_1\left(\hat{I}_{s\beta} - I_{s\beta}\right) \\ \dot{\hat{\phi}}_{r\alpha} = \alpha_4 \hat{I}_{s\alpha} - \alpha_5 \hat{\phi}_{r\alpha} - p\omega\hat{\phi}_{r\beta} + g_3\left(\hat{I}_{s\alpha} - I_{s\alpha}\right) - g_4\left(\hat{I}_{s\beta} - I_{s\beta}\right) \\ \dot{\hat{\phi}}_{r\beta} = \alpha_4 \hat{I}_{s\beta} - \alpha_5 \hat{\phi}_{r\beta} + p\omega\hat{\phi}_{r\beta} + g_4\left(\hat{I}_{s\alpha} - I_{s\alpha}\right) + g_4\left(\hat{I}_{s\beta} - I_{s\beta}\right) \end{cases} \tag{18}$$

The system (18) can be written in the following state form:

$$\dot{\hat{X}} = A(\omega)\hat{X} + BU + G_k e_{is} \tag{19}$$

with:

$$e_{is} = \left[\left(\hat{I}_{s\alpha} - I_{s\alpha}\right)\left(\hat{I}_{s\beta} - I_{s\beta}\right)\right]^T$$

$$A(\omega) = \begin{bmatrix} -\alpha_1 & 0 & \alpha_2 & p\alpha_{2T_r\omega} \\ 0 & -\alpha_1 & -p\alpha_2 T_r\omega & \alpha_2 \\ \alpha_4 & 0 & -\alpha_5 & -p\omega \\ 0 & \alpha_4 & p\omega & -\alpha_5 \end{bmatrix} \tag{20}$$

$$[B] = \begin{bmatrix} \frac{1}{\sigma L_s} & 0 \\ 0 & \frac{1}{\sigma L_s} \\ 0 & 0 \\ 0 & 0 \end{bmatrix}, [G_k] = \begin{bmatrix} g_1 & -g_2 \\ g_2 & g_1 \\ g_3 & -g_4 \\ g_4 & g_3 \end{bmatrix}, U = \left[V_{s\alpha} V_{s\beta}\right]$$

Now let's assume that the velocity ω is an unknown constant parameter. The task is to find an adaptation law that allows us to estimate it. The observer can be written as follows, replacing ω by $\hat{\omega}$:

$$\dot{e} = (A + G_k C)e + (\Delta A)\hat{X} \tag{21}$$

with:

$$\Delta A = A(\hat{\omega})\hat{X} - A(\omega)X = \begin{bmatrix} 0 & 0 & 0 & p\alpha_2 T_r \Delta\omega \\ 0 & 0 & -p\alpha_2 T_r \Delta\omega & 0 \\ 0 & 0 & 0 & -p\Delta\omega \\ 0 & 0 & p\Delta\omega & 0 \end{bmatrix} \tag{22}$$

where:

$$\Delta\omega = \hat{\omega} - \omega, e = \hat{X} - X = \left[e_{I_{s\alpha}} e_{I_{s\beta}} e_{\phi_{r\alpha}} e_{\phi_{r\beta}}\right]^T$$

Now, consider the following Lyapunov function

$$V = e^T e + (\Delta\omega)^2/\lambda \tag{23}$$

Its derivative, Eq. (23), with respect to time is:

$$\dot{V} = e^T\left\{(A + G_k C)^T + (A + G_k C)\right\}e + \hat{X}^T(\Delta A)^T e + e^T(\Delta A)\hat{X} + \frac{2}{\lambda}\Delta\omega\dot{\hat{\omega}} \tag{24}$$

then

$$\dot{V} = e^T\left\{(A + G_k C)^T + (A + G_k C)\right\}e - 2\alpha_2 T_r \Delta\omega\left(e_{I_{s\alpha}}\hat{\phi}_{r\beta} - \hat{\phi}_{r\alpha}e_{I_{s\beta}}\right) + \frac{2}{\lambda}\Delta\omega\dot{\hat{\omega}} \tag{25}$$

where:

$$\hat{X}^T(\Delta A)^T e + e^T(\Delta A)\hat{X} = -2\alpha_2 T_r \Delta\omega\left(e_{I_{s\alpha}}\hat{\phi}_{r\beta} - \hat{\phi}_{r\alpha}e_{I_{s\beta}}\right) \tag{26}$$

From Eq. (25) we can deduce the adaptation law for the estimation of the rotor velocity by equalizing the second term and the third term. We obtain:

$$\dot{\hat{\omega}} = -\lambda\alpha_2 T_r \Delta\omega\left(e_{I_{s\alpha}}\hat{\phi}_{r\beta} - \hat{\phi}_{r\alpha}e_{I_{s\beta}}\right) \tag{27}$$

with:
λ: is a positive constant

4.2.3 Determination of the matrix G_k

The elements of the matrix a $G_k a$ re chosen so that the first term of Eq. (25) is definite negative.

$$\dot{V} = e^T\left\{(A + G_k C)^T + (A + G_k C)\right\}e < 0 \tag{28}$$

To have $\dot{V} < 0$ and taking account the sign of e and $\dot{e}$, we introduce a positive $(k.)$. We obtain elements of the matrix G_k as follows:

$$\begin{cases} g_1 = (1-k)\alpha_1 = (1-k)\left(\dfrac{R_s}{\sigma L_s} + \dfrac{1-\sigma}{\sigma T_r}\right) \\ g_2 = (1-k)\hat{\omega} \\ g_3 = (1-k)(\alpha_3\alpha_5 + \alpha_4) = (1-k)\left(\dfrac{M}{T_r} + \dfrac{1-\sigma}{\sigma M T_r}\right) \\ g_4 = -\hat{\omega}(1-k)p\alpha_3 = -p\hat{\omega}(1-k)\left(\dfrac{1-\sigma}{\sigma M}\right) \end{cases} \tag{29}$$

The adaptation law (27) is established for a constant speed [13, 14], the authors of this adaptation method propose the following PI.

$$\dot{\omega} = k_p\left(e_{I_{s\alpha}}\hat{\phi}_{r\beta} - e_{I_{s\beta}}\hat{\phi}_{r\alpha}\right) + k_i\int\left(e_{I_{s\alpha}}\hat{\phi}_{r\beta} - e_{I_{s\beta}}\hat{\phi}_{r\alpha}\right)dt \tag{30}$$

4.3 Fault detection and isolation algorithm

The core of our fault-tolerant control approach, which integrates with the indirect field-oriented control of the induction machine, is its robust fault detection and isolation (FDI) mechanism. As illustrated in **Figure 5**, this mechanism primarily relies on monitoring the three stator phase currents (I_a, I_b, I_c). The fundamental premise for identifying sensor anomalies stems from the inherent property of a balanced three-phase system:

$$I_a + I_b + I_c = 0 \tag{31}$$

This principle allows us to infer one current if the other two are known. In practice, measured currents will naturally include some measurement noise. From Eq. (31), we can measure two currents and the third one can be calculated as follows.

$$\begin{cases} I_{a,m} = I_a + \eta_a \\ I_{b,m} = I_b + \eta_b \\ I_{c,m} = I_c + \eta_c \end{cases} \tag{32}$$

with, (η_a, η_b, η_c) are measurement noise.

To bolster the fault detection capability, three distinct Kubota observers (Ob_k, for k = 1, 2, 3) are employed. Each observer is strategically configured to operate using only two directly measured phase currents, while the third current's value is derived using the balancing Eq. (31). This redundancy provides independent estimations. The specific current inputs for each observer are therefore defined as:

$$\begin{cases} I_{a1} = I_{a,m}, I_{b1} = I_{b,m}, I_{c1} = -I_{a,m} - I_{c,m} \\ I_{a2} = I_{a,m}, I_{b,2} = -I_{a,m} - I_{c,m}, I_{c,2} = I_{c,m} \\ I_{a,3} = -I_{c,m} - I_{b,m}, I_{b,3} = I_{b,m}, I_{c,3} = I_{c,m} \end{cases} \tag{33}$$

In the coordinate system (α, β), the two current components are given by:

$$\begin{cases} I_{s\alpha,k} = \frac{1}{3}(2I_{a,k} - I_{b,k} - I_{c,k}) \\ I_{s\beta,k} = \frac{\sqrt{3}}{2}(I_{sb,k} - I_{sc,k}) \end{cases} \tag{34}$$

The comprehensive integration of these components within the fault-tolerant control scheme is further detailed in the overall block diagram presented in **Figure 5**.

4.4 Residual generation block

The residual generation block is crucial for fault detection and operates by comparing the estimated and measured components of the rotor flux. The current residuals are noted r_k, $k = 1, 2, 3$, and they are calculated such as:

$$r_k = \left|\hat{\phi}_{\alpha,k} + \phi_{\alpha,k}\right| + \left|\hat{\phi}_{\beta,k} + \phi_{\beta,k}\right| \tag{35}$$

The residue r_k are then processed using a low-pass filter H(s) to remove high-frequency noise and provide smoother fault indicators.

Low pass filter $H(s)$ is defined by:

$$H(s) = \frac{1}{(\tau s + 1)} \tag{36}$$

with:

τ: filter time constant

$\bar{r_k}$: filtered residual vector

4.5 Transition block

The transition block plays a pivotal role in determining the operational status of the sensors (nominal or faulty). Fault detection is achieved by comparing the filtered residuals r_k to a certain detection threshold depending on T, e_{r1}, e_{r2} and e_{r3}.

This detection limit must be such that $T > e_{rk}(t)$ for $k = 1, 2, 3$. Fault detection is then carried out as follows:

$$\begin{cases} |r(t)| \leq T \rightarrow f(t) = 0 \\ |r(t)| > T \rightarrow f(t) \neq 0 \end{cases} \tag{37}$$

The result of the comparison to a threshold is a Boolean quantity, we will symbolically write that the residual $\bar{r_k} = 1$ si $|\bar{r_k}(t)| > T$ else $r_k = 0$.

After the signature of the fault, it is decided which observer Ob_k pour $k = 1, 2, 3$ working. Residual calculation Eq. (38) enables detection of the faulty sensor:

$$\begin{cases} \bar{r_1}(t > t_f) = [100]^T \rightarrow Ob_3(I_{s\alpha 3}, I_{s\beta 3}, \hat{\omega}_3) \\ \bar{r_2}(t > t_f) = [010]^T \rightarrow Ob_2(I_{s\alpha 2}, I_{s\beta 2}, \hat{\omega}_2) \\ \bar{r_3}(t > t_f) = [001]^T \rightarrow Ob_1(I_{s\alpha 1}, I_{s\beta 1}, \hat{\omega}_1) \end{cases} \tag{38}$$

Once a fault signature is identified, a decision is made to select the appropriate, healthy observer (Ob_k for k = 1,2,3) for continued operation. The specific residual patterns, as defined in Eq. (38), enable the isolation of the faulty sensor and activation of the corresponding observer:

5. Hardware setup and implementation

To validate the proposed Fault-Tolerant Control (FTC) strategy, an experimental test bench was constructed. This setup incorporates a low-power induction motor and a real-time control system, and is located at the LTI laboratory (EA 3899) within the IUT of Soissons, France.

The configuration of the experimental system is detailed in **Figure 6**, with its specific parameters provided in **Table 1**. A 1.5 kW induction machine forms the core of the system, controlled by a dSpace 1104 Digital Signal Processor (DSP). This DSP is augmented by an ADMC201 coprocessor, specifically dedicated to induction machine control. The system operates at a fixed sampling frequency of 10 kHz. Stator current measurements are acquired by the controller *via* two 8-bit Analog-to-Digital (A/D) converters. Subsequently, using Pulse Width Modulation (PWM) techniques, the calculated reference voltages are transmitted to the machine through a voltage-source inverter, which also operates at a switching frequency of 10 kHz (**Figure 6**).

Figure 6.
Hardware setup for experimental verification of fault-tolerant control strategy of an IM.

R_s: 5.72 Ω	stator resistance
R_r: 4.2 Ω	rotor resistance rotor
L_s: 0.462 H	stator inductance
L_r: 0.462 H	rotor inductance
M: 0.44 H	mutual inductance
J: 0.0049 Kgm^2	moment of inertia
P: 2	Number of pair of poles
f: 0.003 Nm.s/rad	damping coefficient
Power	1.5 Kw

Table 1.
Induction machine parameters.

6. Discussion of experimental results

In this section, we intentionally introduced a fault by disconnecting the current sensor in phase '**a**' during motor operation. This test is a realistic fault scenario, allowing us to evaluate the efficacy of our Fault-Tolerant Control (FTC) system. The primary goal of the FTC system in such an event is threefold: to detect the fault, isolate the faulty sensor, and subsequently maintain stable motor control by utilizing the remaining healthy sensors in conjunction with the Kubota observers.

Figure 7 illustrates the current response when a fault occurs in phase 'a', clearly indicating the moment of sensor failure. **Figure 8** demonstrates the effectiveness of the residual generation and isolation mechanism. The plots of residual variables, Res1, Res2, and Res3, distinctly show how this particular fault (phase '**a**' sensor disconnection) is precisely isolated.

A significant deviation in a specific residual, while others remain stable, serves as the unique signature of the fault location. This confirms the FDI algorithm's ability to not only detect a problem but also pinpoint its origin. **Figure 9** presents the motor speed response, a critical indicator of system performance under fault conditions. **Figure 7a** explicitly marks the time at which the current sensor fault occurs. **Figure 9** then showcases the motor speed response both before and after the simulated offset fault. Immediately following the fault, a small deviation or ripple in the motor speed is observable. This momentary disturbance signifies the system's initial reaction to the uncompensated fault. However, a key observation here is the rapid and effective fault compensation enacted by the FTC system. This compensation successfully removes the speed ripple, demonstrating the system's ability to recover and restore near-nominal motor speed, thereby ensuring continued, stable operation despite the sensor failure. This highlights the successful transition and re-configuration of the control strategy post-fault detection.

7. Conclusion

This work successfully proposed and experimentally validated a robust sensor fault-tolerant vector control strategy specifically designed for induction motor drives. By integrating three redundant current sensors and their associated Kubota flux observers directly within the Field-Oriented Control (FOC) framework, the developed system demonstrably maintained critical control performance even when faced

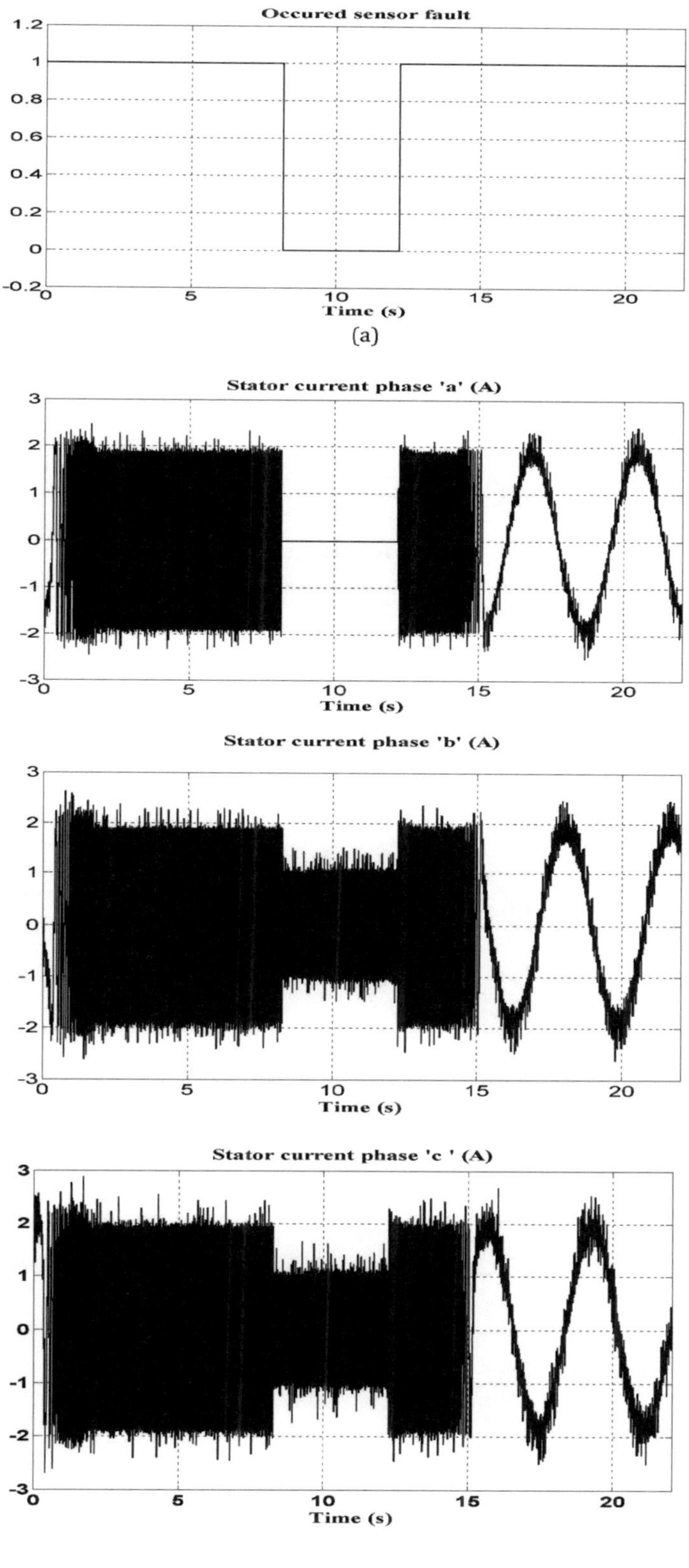

Figure 7.
Responses of the drive and the IFDIS for the phase "i_a" sensor fault.

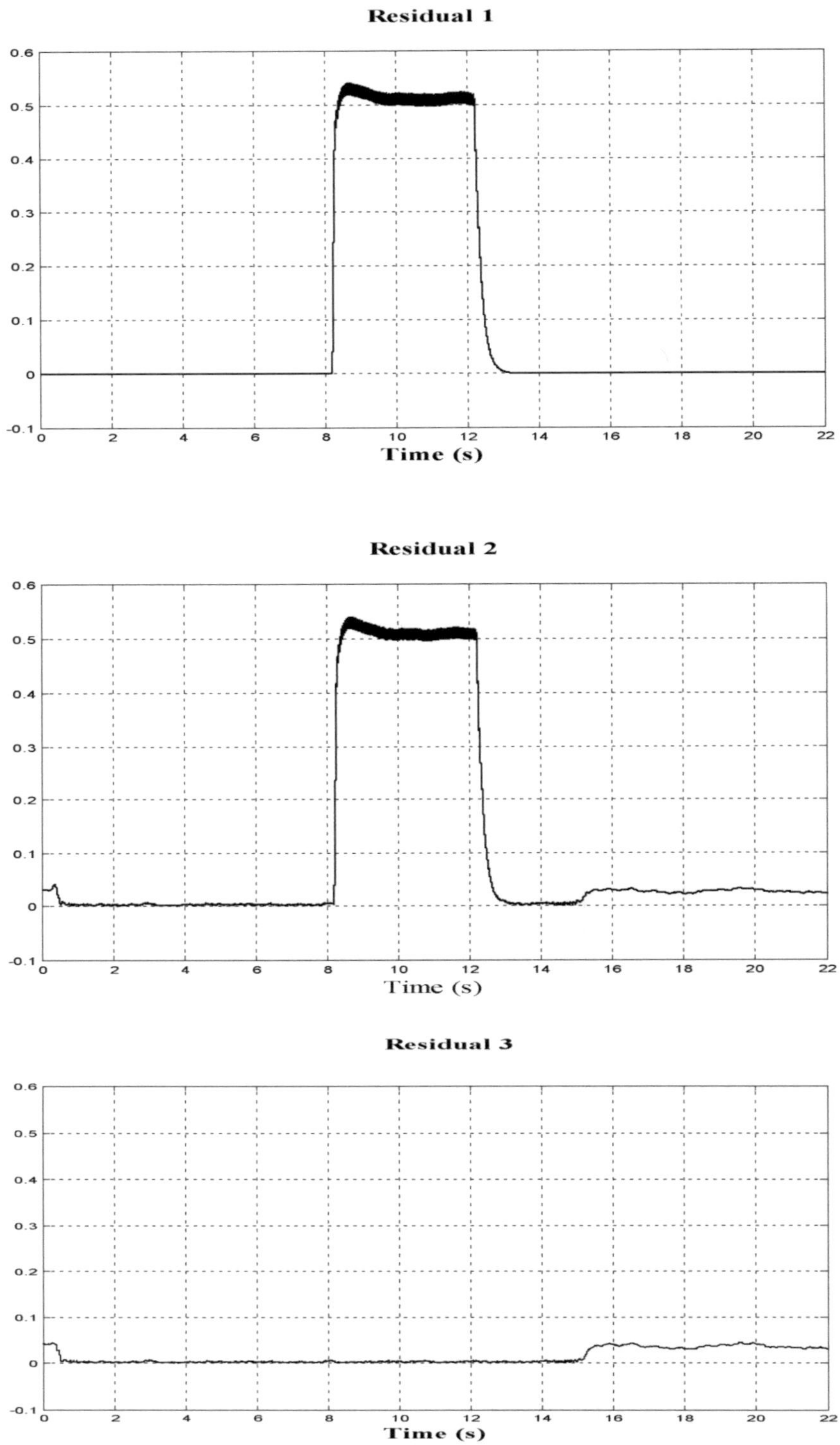

Figure 8.
Responses of residual variables to sensor fault.

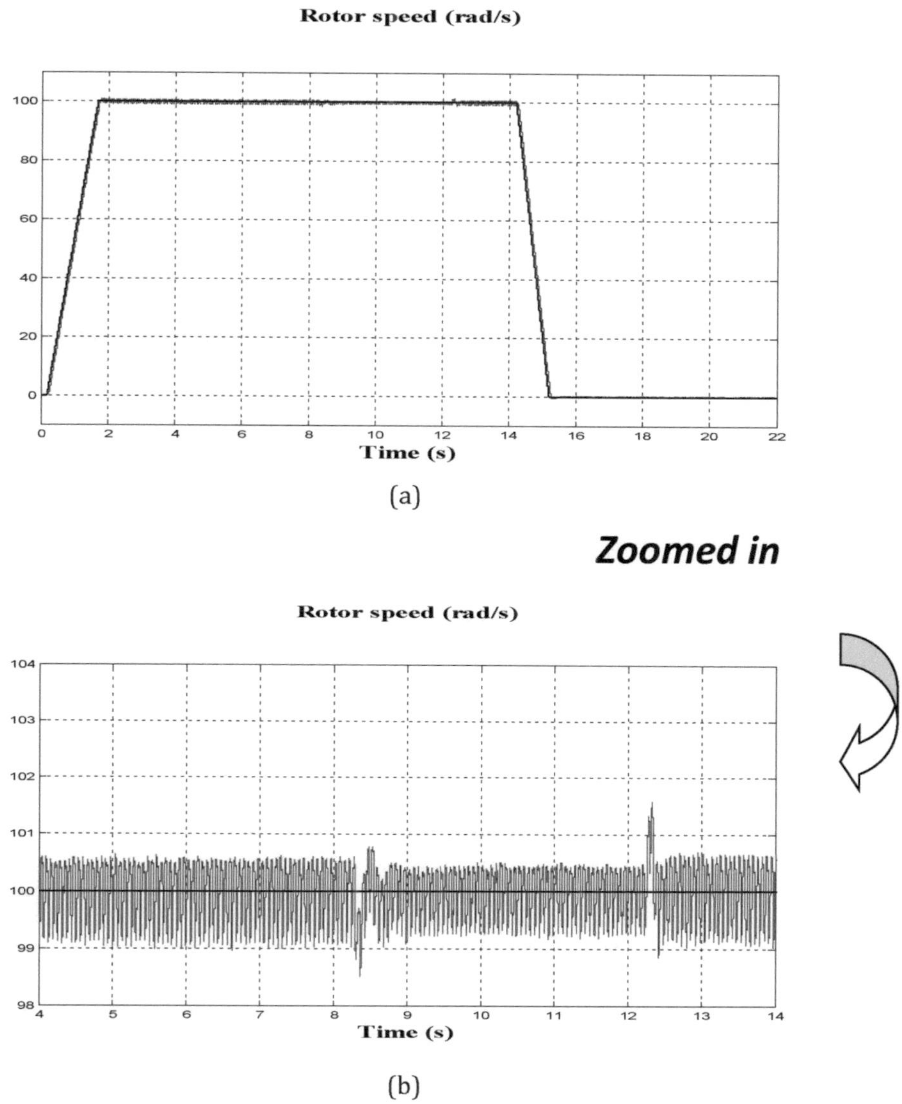

Figure 9.
(a) Speed response to the fault-tolerant control of induction machine. (b) speed response in steady state (Zoomed in).

with sensor fault conditions. The compelling experimental results unequivocally confirmed the effectiveness of the proposed approach, showing its capability to not only successfully isolate the faulty sensor but also to seamlessly compensate for its impact. This significantly enhances overall system reliability and ensures continuous, uninterrupted operation, which is paramount in industrial applications.

Acknowledgements

The author acknowledges the assistance of ChatGPT (OpenAI) for language polishing of the manuscript.

Author details

Fatiha Zidani
LSPIE Laboratory "Laboratory of Electromagnetic Induction and Propulsion Systems", Department of Electrical Engineering, Faculty of Technology, Batna-2 University, Algeria

*Address all correspondence to: f.zidani@univ-batna2.dz

References

[1] Choqueuse V, Benbouzid MEH, Amirat Y, Turri S. Diagnosis of three-phase electrical machines using multidimensional demodulation techniques. IEEE Transactions on Industrial Electronics;**59**(4):2014-2023

[2] Amin AA, Hasan KM. A review of fault tolerant control systems: Advancements and applications. Elsevier Journal. 2019;**143**:58-68

[3] Demir R. Robust stator flux and load torque estimations for induction motor drives with EKF-based observer. Electrical Engineering. 2023;**105**:551-562

[4] Choudhary A, Goyal D, Shimi SL, et al. Condition monitoring and fault diagnosis of induction motors: A review. Archives of Computational Methods in Engineering. 2019;**26**(4):1221-1238

[5] Romeo ME, Seron MM. Switching strategy for sensor fault tolerant vector control of doubly fed induction machines. In: Proceedings of the 2nd Conference on Control and Fault-Tolerant Systems (SysTol), Nice, France, Oct. 6–8, 2010. Piscataway, NJ, USA: IEEE; 2010. pp. 209-214. ISBN: 978-1-4244-8152-8

[6] Tahami F, Shojaei A, Khatir DA. A diversity-based reconfigurable method for fault tolerant control of induction motors. In: 2006 International Symposium on Power Electronics, Electrical Drives, Automation and Motion (SPEEDAM), Taormina, Italy, May 23-26, 2006. Piscataway, NJ, USA. ISBN 1-4244-0193-3: IEEE. pp. 12-17. DOI: 10.1109/SPEEDAM11060.2006

[7] Diallo D, Benbouzid MEH, Makouf A. A fault-tolerant control architecture for induction motor drives in automotive applications. IEEE Transactions on Vehicular Technology. 2004;**53**(6): 1847-1855

[8] Benbouzid MEH, Diallo D, Zeraoulia M. Advanced fault-tolerant control of induction-motor drives for EV/HEV traction applications: From conventional to modern and intelligent control techniques. IEEE Transactions on Vehicular Technology. 2007;**56**(2):519-528

[9] Robyns B, François B, Degobert P, Hautier J-P. Vector Control of Induction Machines: Desensitisation and Optimisation through Fuzzy Logic. London: Springer; 2012. pp. 75-121

[10] Romero M, Seron M, de Doná J. Sensor fault tolerant control of induction motors. IFAC Proceedings. 2008;**41**(2): 1230-1235

[11] Zidani F, Benbouzid MEH, Diallo D, Nait-Said MS. Induction motor stator faults diagnosis by a current Concordia pattern-based fuzzy decision system. IEEE Transactions on Energy Conversion. 2003;**18**(4):469-475

[12] Zidani F, Diallo D, Benbouzid MEH, Nait Said R. Fuzzy detection and diagnosis of fault modes in a voltage-PWM inverter induction motor drive. IEEE Transactions on Industrial Electronics. 2008;**55**(2):586-593

[13] Kubota H, Matsuse K, Nakano T. DSP-based speed adaptive flux observer of induction motor. IEEE Transactions on Industry Applications. 1993;**29**(2): 344-348

[14] Kubota H, Matsuse K. Speed sensorless field-oriented control of induction motor with rotor resistance adaptation. IEEE Transactions on Industry Applications. 1994;**30**(5): 1219-1224

Chapter 3

The Set of Functions Corresponding to First-Order Chemical Kinetics Problems, Its Structure, and Related Inverse Problems

Victor Martinez-Luaces

Abstract

First-order chemical kinetics can be considered—at least mathematically—the most common ones since, under certain conditions, many other types of reactions also give a similar mathematical model, that is, a linear ODE system. In this chapter, we propose to study the set of functions that can be solutions to first-order chemical kinetics problems. First, we observe that this set of functions cannot have a vector space structure, although it is possible to define a norm. Also, the stability of the solutions is proved, and the importance of this result in chemistry is analyzed. Second, we move on to the central core of the proposal: the related inverse problems. This set of functions gives rise to several interesting inverse questions. For example: 1) Existence: Given a function of this set, is there always a mechanism that has it as a solution? 2) Uniqueness: Can two different mechanisms give the same solution? 3) Stability: Do small changes in the norm of the function always produce small changes in the corresponding mechanism? The chapter ends with a relevant conjecture and two open questions about other inverse problems.

Keywords: chemical kinetics functions, associated matrices, stability of ODE systems, inverse problems, existence and uniqueness, eigenvalues and eigenvectors

1. Introduction

A first-order chemical reaction (FOCR) occurs when a chemical species S_1 gives rise to one or more products (P_1, P_2, etc.), such that the reaction rate is proportional to $[S_1]$, the remaining concentration of the species S_1, that acts as the reactant (**Figure 1**).

The mathematical model corresponding to this FOCR is the following:

$$\frac{d[S_1]}{dt} = -k[S_1], \tag{1}$$

where $[S_1]$ is the concentration of the reactant S_1, t is time, and k is the proportionality constant, called the kinetic constant of the FOCR.

Figure 1.
First-order chemical reaction (FOCR).

When several FOCRs occur involving the same chemical species, then we have a first-order chemical kinetics mechanism (FOCKM). We presented several examples of FOCKMs in previous works (e.g., [1, 2]). In those works, we studied in which cases the associated matrix of the ODE system is a lower triangular one, the signs of the eigenvalues, their algebraic and geometric multiplicities, and other related topics.

In this work, we study the set of functions of all possible theoretical solutions of FOCKM problems. We observe that for this set it is impossible to have a vector space structure; however, it is possible to define an appropriate norm. Also, we prove the stability of the solutions, and we discuss the importance of this result in chemistry.

Finally, we analyze several inverse problems related to the set of functions mentioned above, and we propose some open questions relevant for further research on this topic.

2. An illustrative example of a first-order mechanism

An interesting case, where a FOCKM involving five chemical species, is presented in **Figure 2** (Example N° 1). In fact, **Figure 2a** shows the reactions and the numerical values of the kinetic constants, while **Figure 2b** schematizes the same FOCKM as a directed graph.

The reactant S_1 gives the product S_2 by means of an FOCR whose kinetic constant is $k_{12} = 4$, and so, the corresponding ODE is:

$$\frac{d[S_1]}{dt} = -4[S_1]. \tag{2}$$

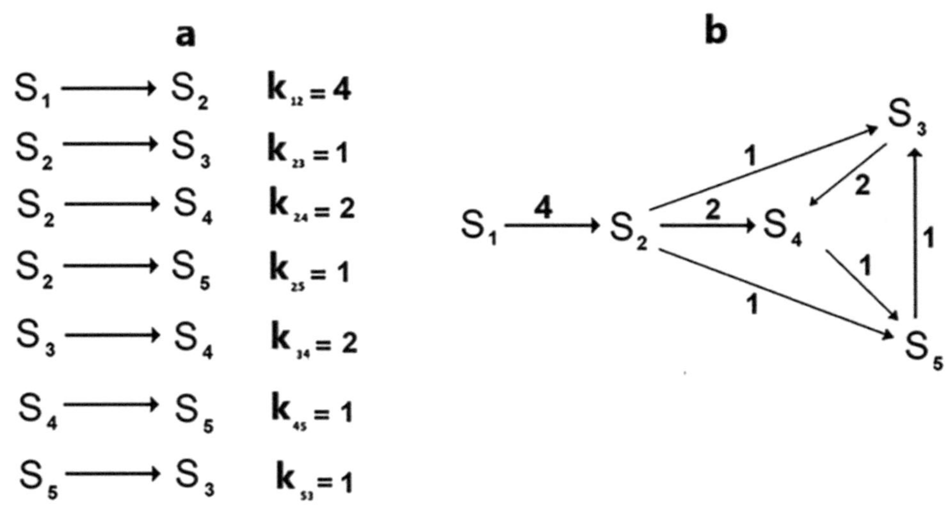

Figure 2.
First-order chemical kinetics mechanism (FOCKM).

For the species S_2 the situation is somewhat more complex, since it is obtained from the reactant S_1, by means of the aforementioned FOCR, and at the same time, it is consumed by three different reactions of kinetic constants: $k_{23} = 1$, $k_{24} = 2$, and $k_{25} = 1$. Therefore, the corresponding ODE is

$$\frac{d[S_2]}{dt} = 4[S_1] - 1[S_2] - 2[S_2] - 1[S_2] = 4[S_1] - 4[S_2]. \tag{3}$$

Proceeding in the same way, the other three ODEs of the system are obtained:

$$\frac{d[S_3]}{dt} = 1[S_2] - 2[S_3] + 1[S_5]. \tag{4}$$

$$\frac{d[S_4]}{dt} = 2[S_2] + 2[S_3] - 1[S_4]. \tag{5}$$

$$\frac{d[S_5]}{dt} = 1[S_2] + 1[S_4] - 1[S_5]. \tag{6}$$

Eqs. (2)–(6) give an ODE system that can be written as

$$\begin{cases} \dfrac{d[S_1]}{dt} = -4[S_1] \\ \dfrac{d[S_2]}{dt} = 4[S_1] - 4[S_2] \\ \dfrac{d[S_3]}{dt} = [S_2] - 2[S_3] + [S_5] \\ \dfrac{d[S_4]}{dt} = 2[S_2] + 2[S_3] - [S_4] \\ \dfrac{d[S_5]}{dt} = [S_2] + [S_4] - [S_5] \end{cases} \tag{7}$$

And the corresponding associated matrix is:

$$\mathbb{A} = \begin{pmatrix} -6 & 0 & 0 & 0 & 0 \\ 1 & -2 & 0 & 0 & 0 \\ 2 & 2 & -1 & 0 & 0 \\ 1 & 0 & 1 & -1 & 0 \\ 2 & 0 & 0 & 0 & 0 \end{pmatrix}. \tag{8}$$

We will call this matrix—associated with the previous ODE system—a FOCKM matrix, and later we will study its structure and some of its properties.

The Jordan canonical form [3] corresponding to this matrix is given by

$$\mathbb{J} = \begin{pmatrix} \mathbb{B}_1 & \mathbb{O} & \mathbb{O} \\ \mathbb{O} & \mathbb{B}_2 & \mathbb{O} \\ \mathbb{O} & \mathbb{O} & \mathbb{B}_3 \end{pmatrix}, \tag{9}$$

where the diagonal blocks are

$$\mathbb{B}_1 = (0), \mathbb{B}_2 = \begin{pmatrix} -4 & 1 \\ 0 & -4 \end{pmatrix}, \text{and } \mathbb{B}_3 = \begin{pmatrix} -2-i & 0 \\ 0 & -2+i \end{pmatrix}. \quad (10)$$

Therefore, the system solutions (i.e., the concentrations of chemical species as a function of time) are linear combinations of the following functions: $\{e^{0t}, e^{-4t}, te^{-4t}, e^{-2t}\cos t, e^{-2t}\sin t\}$, e.g., [4, 5]. In other words, the solutions to the FOCKM problem are of the form

$$[S_i](t) = C_1 + C_2\, e^{-4t} + C_3\, te^{-4t} + C_4 e^{-2t}\cos t + C_5\, e^{-2t}\sin t \quad (11)$$

Inverse problems regarding the FOCKM problem solutions are analyzed in the next sections.

3. Previous results about FOCKM problems revisited

If n chemical species, $S_1, S_2, \dots, S_n$ are considered and we use the same notation as in Example N°1, then the reactions between these species can be written as.

$S_i \overset{k_{ij}}{\rightarrow} S_j$, where the reaction is an FOCR and k_{ij} is the corresponding kinetic constant. The corresponding ODE system—constructed as in Example 1—is the following:

$$\frac{d[S_1]}{dt} = -(k_{12} + k_{13} + \dots + k_{1n})[S_1] + k_{21}[S_2] + k_{31}[S_3] + \dots + k_{n1}[S_n] \quad (12)$$

$$\frac{d[S_2]}{dt} = k_{12}[S_1] - (k_{21} + k_{23} + \dots + k_{2n})[S_2] + k_{32}[S_3] + \dots + k_{n2}[S_n] \quad (13)$$

…

$$\frac{d[S_n]}{dt} = k_{1n}[S_1] + k_{2n}[S_2] + \dots + k_{n-1n}[S_{n-1}] - (k_{n1} + k_{n2} + \dots + k_{nn-1})[S_n] \quad (14)$$

And the FOCKM matrix can be written as

$$\mathbb{A} = \begin{pmatrix} -(k_{12} + k_{13} + \dots + k_{1n}) & \cdots & k_{n1} \\ \vdots & \ddots & \vdots \\ k_{1n} & \cdots & -(k_{n1} + k_{n2} + \dots + k_{nn-1}) \end{pmatrix} \quad (15)$$

This is absolutely general if we add the convention that in case there is no reaction, the kinetic constant is zero (see, e.g., [1]).

Consequently, the next definitions arise:

Definition 1.

$\mathbb{A}$ is a FOCKM matrix if it has the following structure:

$$\mathbb{A} = \begin{pmatrix} -\sigma_1 & \cdots & a_{1,n} \\ \vdots & \ddots & \vdots \\ a_{n,1} & \cdots & -\sigma_n \end{pmatrix}, \quad (16)$$

where: $\sigma_1 = \sum_{i \neq 1} a_{i,1}, \dots,$

$$\sigma_n = \sum_{i \neq n} a_{i,n}. \quad (17)$$

Note that the entries $a_{i,j}$ are kinetic constants $k_{j,i}$—or zero if there is no reaction—so $a_{i,j} \geq 0$ for all $i \neq j$.

Definition 2.

A linear ODE system $\mathbb{X}' = \mathbb{A}\,\mathbb{X}$, where $\mathbb{A}$ is a FOCKM matrix, will be called the FOCKM problem.

If the rows of a FOCKM matrix are added, the result is always the null vector, due to its particular structure. From this observation, it is easy to prove the following result:

Previous Result N° 1.

If $\mathbb{A}$ is a FOCKM matrix with $n \geq 2$ rows, then $\lambda = 0$ is an eigenvalue of $\mathbb{A}$.

Now, if the Gershgorin's Circles Theorem [6] is applied to the columns of $\mathbb{A}$, a given FOCKM matrix, then it is easy to prove this interesting result:

Previous Result N° 2.

If $\mathbb{A}$ is a FOCKM matrix with $n \geq 2$ rows, then all its eigenvalues lie in a disk $\mathbb{D}((-r, 0), r)$ with $r > 0$.

The details of the proof can be read in [1]. In particular, it can be observed that the minimum value of r is $r = Max\ \{\sigma_1, \sigma_2, \ldots, \sigma_n\}$.

Note that if $r > 0$ the disk centered on $(-r, 0)$, with radius r looks like the one in **Figure 3**.

As usually happens in linear ODE systems, the solutions are linear combinations of exponentials, oscillatory functions (sines and cosines), polynomials, and eventually, products of these functions, as in Example N°1. However, there are some solutions that are not possible:

1. Purely oscillatory solutions, like $[S_i](t) = C_1 \cos(bt) + C_2\ \sin(bt)$ cannot occur, since this would require eigenvalues of the form $\lambda = 0 \pm ib$, with $b \neq 0$, and this fact contradicts previous result N° 2.

2. Oscillatory-polynomial solutions, like $[S_i](t) = C_1 t \cos(bt) + C_2 t \sin(bt)$ cannot occur for the same reasons (see for example [2]).

3. Purely polynomial solutions, like $[S_i](t) = C_1 + C_2 t + \ldots + C_p t^p$ cannot occur either. This fact has been chemically justified (see [2]), but there also exists an algebraic proof that supports this conclusion (see [1]). The central core of this proof is based on the next theorem:

Previous Result N° 3.

If $\mathbb{A}$ is a FOCKM matrix with $n \geq 2$ rows, the algebraic and geometric multiplicities of the null eigenvalue coincide, that is: $AM_{\lambda=0} = GM_{\lambda=0}$.

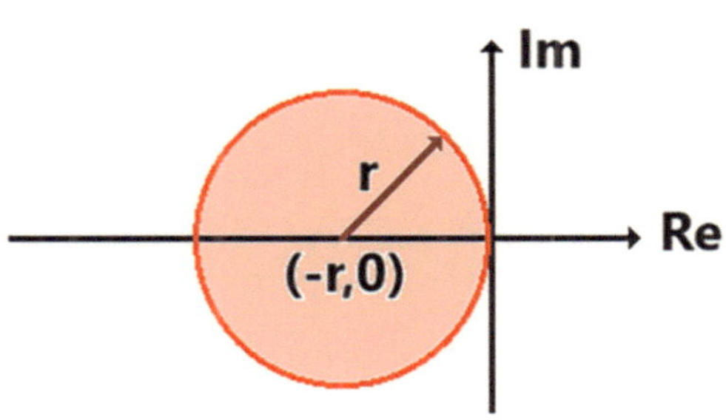

Figure 3.
Diagram of the union of the Gershgorin's disks.

Details of the proof can be read in [1].

Taking into account the previous results, the only "pure" case that can take place is the combination of exponentials, including e^{0t}, and considering the "mixed" cases, in a first approach only the oscillatory-polynomial can be discarded.

4. Fundamental definitions

In Example N° 1, only two "pure" exponentials appear: e^{0t} and e^{-4t}. It is worth mentioning that exponentials with a negative or zero exponent are a direct consequence of the previous result, N° 2; therefore, it is convenient to make the next definition:

Definition 3.

Let the set of exponentials, $E_{\leq 0} = \{e^{\lambda t}, \lambda \leq 0\}$, with a non-positive exponent.

Once again, in Example N° 1 other functions appear, like te^{-4t}, $e^{-2t}\cos t$, and $e^{-2t}\sin t$, so we will consider functions that are products of exponentials, polynomials, and oscillatory functions (EPO).

Definition 4.

Let be the set of functions.

$EPO = \{e^{at}p(t)\ cos\,(bt) : a<0, b\in\mathbb{R}, p(t)\ polynomial\} \cup \{e^{at}p(t)\ sin\,(bt) : a<0\}$, $b\in\mathbb{R}, p(t)\ polynomial\}$, which includes products of negative exponentials, polynomials, and oscillations (sines and cosines).

And now we consider the set of functions that are linear combinations of those that belong to the previous sets, as can be seen in the following definition:

Definition 5.

Let $\mathcal{L}$ be the following set of functions:

$$\mathcal{L} = \{f = \alpha_1 f_1 + \alpha_2 f_2 + \cdots + \alpha_n f_n : n \geq 1, f_i \in E_{\leq 0} \cup EPO, \forall i = 1 \dots n\}$$

It should be noted that the last definition of this section refers to the set of functions that is the object of study of the chapter.

Definition 6.

Let $\mathcal{F}_{FOCKM}$ be the following set of functions:

$\mathcal{F}_{FOCKM} = \{f : f(t)\ is\ a\ solution\ of\ a\ FOCKM\ problem\}$.

This means that $f(t)$ is a solution of an ODE system $\mathbb{X}' = \mathbb{A}\,\mathbb{X}$, where $\mathbb{A}$ is a FOCKM matrix. The aim of this chapter is to study whether this set of functions can have a vector space structure, if it is possible to define a norm, the stability of the solutions, and some related inverse problems.

5. About the set of functions $\mathcal{F}_{FOCKM}$

Taking into account what has been analyzed in sections 2 and 3, it is clear that $\mathcal{F}_{FOCKM} \subseteq \mathcal{L}$, so the first obvious question is whether equality can take place and under what conditions.

Question N° 1: Is it possible that $\mathcal{F}_{FOCKM} = \mathcal{L}$?

The answer to this question is no. If $f \in \mathcal{F}_{FOCKM}$, then $f(t)$ represents the concentration of a certain chemical species, and it cannot be negative. Thus, for example, if $f(t) = -e^{-t}$ then $f \notin \mathcal{F}_{FOCKM}$ although $f \in \mathcal{L}$.

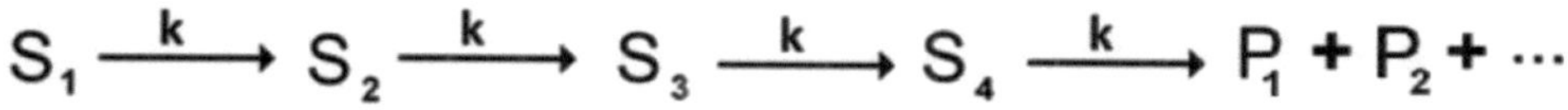

Figure 4.
Three consecutive reactions.

Returning once more to Example N° 1, we saw that the solutions were of the form

$$[S_i](t) = C_1 + C_2\, e^{-4t} + C_3\, te^{-4t} + C_4 e^{-2t} \cos t + C_5\, e^{-2t} \sin t \tag{18}$$

so $\{e^{-4t}, te^{-4t}\} \subseteq \mathcal{F}_{FOCKM}$, and it is immediate to ask if other functions such as t^2e^{-4t}, t^3e^{-4t}, etc., are also part of this set.

Question N° 2: Is it possible to obtain as t^2e^{-4t}, t^3e^{-4t}, etc., as solutions for a given FOCKM problem?

The answer to this question is affirmative, as shown in Example N° 2, schematized in **Figure 4**.

The ODE system and the associated matrix are the following:

$$\begin{cases} \dfrac{d[S_1]}{dt} = -k[S_1] \\ \dfrac{d[S_2]}{dt} = k[S_1] - k[S_2] \\ \dfrac{d[S_3]}{dt} = k[S_2] - k[S_3] \\ \dfrac{d[S_4]}{dt} = k[S_3] - k[S_4] \end{cases}, \tag{19}$$

and

$$\mathbb{A} = \begin{pmatrix} -k & 0 & 0 & 0 \\ k & -k & 0 & 0 \\ 0 & k & -k & 0 \\ 0 & 0 & k & -k \end{pmatrix} \tag{20}$$

In this case the corresponding Jordan matrix is:

$$\mathbb{J} = \begin{pmatrix} 0 & 0 & 0 & 0 \\ 0 & -k & 1 & 0 \\ 0 & 0 & -k & 1 \\ 0 & 0 & 0 & -k \end{pmatrix} \tag{21}$$

Hence, the solutions are linear combinations of $\{e^{0t}, e^{-kt}, te^{-kt}, t^2e^{-kt}\}$.

Following the same procedure, we can add more reactions with the same kinetic constant k, like in **Figure 5**.

In this case, other solutions of the form $f_p(t) = t^p e^{-kt}$ are obtained, so the answer to Question N° 2 is affirmative.

Question N° 3: Is it possible to obtain solutions of the form $te^{-at} \cos(bt)$ and $te^{-at} \sin(bt)$ for a given FOCKM problem?

Once again, the answer is affirmative (**Figure 6**).

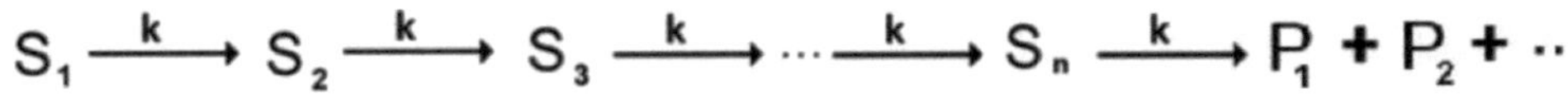

Figure 5.
FOCKM with n consecutive reactions.

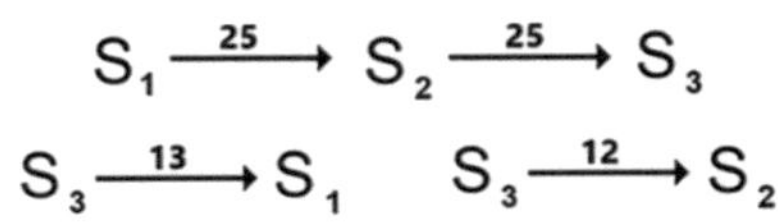

Figure 6.
FOCKM with four reactions and three chemical species.

In order to see this result, let us consider first the FOCKM matrix $\mathbb{F}$, where

$$\mathbb{F} = \begin{pmatrix} -25 & 0 & 13 \\ 25 & -25 & 12 \\ 0 & 25 & -25 \end{pmatrix}, \tag{22}$$

which corresponds to the mechanism of **Figure 6**.

Second, let us consider the non-FOCKM matrix $\mathbb{N}$, being

$$\mathbb{N} = \begin{pmatrix} -30 & 0 & 15 \\ 15 & -30 & 0 \\ 0 & 15 & -30 \end{pmatrix}. \tag{23}$$

It should be noted that adding all the rows of $\mathbb{N}$ the result is: $(-,15, \quad -,15, \quad -,15,) \neq \vec{0}$.

Finally, we consider the diagonal matrix:

$$\mathbb{D} = \begin{pmatrix} 15 & 0 & 0 \\ 0 & 15 & 0 \\ 0 & 0 & 15 \end{pmatrix}. \tag{24}$$

If we use all these matrices as blocks of the 6×6 matrix $\mathbb{A}$, we have

$$\mathbb{A} = \begin{pmatrix} \mathbb{F} & \mathbb{D} \\ \mathbb{O} & \mathbb{N} \end{pmatrix}, \tag{25}$$

with $\mathbb{O}$ being the null 3×3 matrix.

It is easy to observe that $\mathbb{A}$ is a FOCKM matrix, and its eigenvalues are: $\lambda_1 = 0$, $\lambda_2 = -15$ and $\lambda_{3,4} = \frac{\pm 15\, i\, \sqrt{3}-75}{2}$.

Moreover, its corresponding Jordan matrix is:

$$\mathbb{J} = \begin{pmatrix} 0 & 0 & 0 & 0 & 0 & 0 \\ 0 & -15 & 0 & 0 & 0 & 0 \\ 0 & 0 & \lambda_3 & 1 & 0 & 0 \\ 0 & 0 & 0 & \lambda_3 & 0 & 0 \\ 0 & 0 & 0 & 0 & \lambda_4 & 1 \\ 0 & 0 & 0 & 0 & 0 & \lambda_4 \end{pmatrix}. \tag{26}$$

Then, solutions of the form $te^{-\frac{75}{2}t}\cos\left(\frac{15\sqrt{3}}{2}t\right)$ and $te^{-\frac{75}{2}t}\sin\left(\frac{15\sqrt{3}}{2}t\right)$ appear when solving the corresponding ODE system.

Question N° 4: Is it possible to obtain solutions of the form $t^k e^{-at}\cos(bt)$ and $t^k e^{-at}\sin(bt)$ with $k > 1$ for a given FOCKM problem?

In order to answer this question, we will consider the following matrix:

$$\mathbb{A} = \begin{pmatrix} \mathbb{F} & \mathbb{D} & \mathbb{O} & \mathbb{O} \\ \mathbb{O} & \mathbb{N} & \mathbb{D} & \mathbb{O} \\ \mathbb{O} & \mathbb{O} & \mathbb{N} & \mathbb{D} \\ \mathbb{O} & \mathbb{O} & \mathbb{O} & \mathbb{N} \end{pmatrix}, \tag{27}$$

where $\mathbb{F}$, $\mathbb{N}$, and $\mathbb{D}$ are given by (22), (23) and (24), respectively, and $\mathbb{O}$ is the null 3×3 matrix.

It is easy to see that $\mathbb{A}$ is a 12×12 FOCKM matrix, and its corresponding Jordan form is the following:

$$\mathbb{J} = \begin{pmatrix} \mathbb{B}_1 & \mathbb{O} & \mathbb{O} \\ \mathbb{O} & \mathbb{B}_2 & \mathbb{O} \\ \mathbb{O} & \mathbb{O} & \mathbb{B}_3 \end{pmatrix}, \tag{28}$$

where the diagonal blocks are:

$$\mathbb{B}_1 = \begin{pmatrix} 0 & 0 & 0 & 0 \\ 0 & -15 & 1 & 0 \\ 0 & 0 & -15 & 1 \\ 0 & 0 & 0 & -15 \end{pmatrix}, \tag{29}$$

$$\mathbb{B}_2 = \begin{pmatrix} \lambda & 1 & 0 & 0 \\ 0 & \lambda & 1 & 0 \\ 0 & 0 & \lambda & 1 \\ 0 & 0 & 0 & \lambda \end{pmatrix}, \tag{30}$$

$$\mathbb{B}_3 = \overline{\mathbb{B}_2}, \tag{31}$$

the complex conjugate matrix of $\mathbb{B}_2$, and being $\lambda = \frac{15\,i\,\sqrt{3}-75}{2}$.

Then, in this case, solutions of the form $t^k e^{-at}\cos(bt)$ and $t^k e^{-at}\sin(bt)$ with $k = 0, 1, 2, 3$ appear, among others.

Repeating the same procedure with more matrices $\mathbb{F}$, $\mathbb{N}$, and $\mathbb{D}$—given by (22)–(24)—solutions of the form $t^k e^{-at}\cos(bt)$ and $t^k e^{-at}\sin(bt)$ with any $k > 1$, can be obtained.

Taking into account the answers to questions 1, 2, 3, and 4, all the following functions are possible: $e^{0t}, e^{-kt}, t^p e^{-kt}, e^{-at}\cos(bt), e^{-at}\sin(bt)$, $t^p e^{-at}\cos(bt), t^p e^{-at}\sin(bt)$, and also their linear combinations.

6. Norm and boundedness in the set of functions $\mathcal{F}_{FOCKM}$

We observed in the previous sections that if $f \in \mathcal{F}_{FOCKM}$, then $f(t)$ is a linear combination of functions like: $e^{0t}, e^{-kt}, t^p e^{-kt}, e^{-at}\cos(bt), e^{-at}\sin(bt)$,

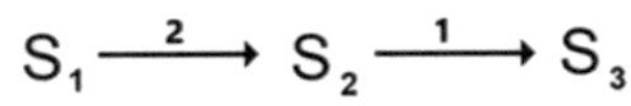

Figure 7.
FOCKM with three reactions and three chemical species.

$t^p e^{-at}\cos(bt), t^p e^{-at}\sin(bt)$, and it is easy to see that all of them except e^{0t} tend to zero when $t \to +\infty$, since $k > 0$ and $a > 0$.

As a consequence, if $f \in \mathcal{F}_{FOCKM}$, then $f(t)$ is bounded for $t \in [0, +\infty)$, and it is possible to define a norm in $\mathcal{F}_{FOCKM}$ as follows:

$$\|f\|_{\infty,[0,+\infty)} = \text{Sup}\{f(t) : t \in [0, +\infty)\} \tag{32}$$

The question that naturally arises is whether it is really a supreme or if it is a maximum. In order to answer this question, let us consider the FOCKM in **Figure 7** (Example N°3).

The corresponding ODE system and its associated matrix are

$$\begin{cases} \dfrac{d[S_1]}{dt} = -2[S_1] \\ \dfrac{d[S_2]}{dt} = 2[S_1] - [S_2] \\ \dfrac{d[S_3]}{dt} = [S_2] \end{cases}, \tag{33}$$

and

$$\mathbb{A} = \begin{pmatrix} -2 & 0 & 0 \\ 2 & -1 & 0 \\ 0 & 1 & 0 \end{pmatrix} \tag{34}$$

And let us suppose that the initial concentrations of the chemical species are: $[S_1](0) = 1$, $[S_2](0) = 0$, and $[S_3](0) = 0$.

Then, solving the system, the result is

$$[S_1](t) = e^{-2t}, [S_2](t) = 2e^{-t} - 2e^{-2t}, \text{and } [S_3](t) = 1 - 2e^{-t} + e^{-2t}, \tag{35}$$

and so, for the species S_1 and S_2 it is a maximum, and for the species S_3 it is a supreme.

Since it must be a general definition, then formula (32) is the correct one.

7. Stability of the solutions and applications in chemistry

It is well known that in a linear ODE system, if all the matrix eigenvalues have a negative real part, then the solutions are stable, and moreover, they are asymptotically stable [7].

In the case of the null eigenvalue, since $AM_{\lambda=0} = GM_{\lambda=0}$ (Previous Result N°3), then the corresponding solutions are stable.

As a consequence, in a FOCKM problem, the solutions are always stable or asymptotically stable. For instance, if we go back to Example N° 3, it is easy to see that the solutions corresponding to the species S_1 are asymptotically stable. Indeed, let $f(t) = c_1 e^{-2t}$ be the solution for species S_1 and let $\tilde{f}(t) = \tilde{c_1} e^{-2t}$ be the solution for slightly different initial conditions; then:

$$\left|f(t) - \tilde{f}(t)\right| = \left|c_1 e^{-2t} - \tilde{c_1} e^{-2t}\right| = |c_1 - \tilde{c_1}| e^{-2t} \underset{t \to \infty}{\to} 0 \tag{36}$$

The same happens for the species S_2 . In this case the solution is

$$g(t) = (2c_1 + c_2)e^{-t} - 2c_1 e^{-2t} = k_1 e^{-t} + k_2 e^{-2t}, \tag{37}$$

and so:

$$|g(t) - \tilde{g}(t)| = \left|\left(k_1 e^{-t} + k_2 e^{-2t}\right) - \left(\tilde{k_1} e^{-t} + \tilde{k_2} e^{-2t}\right)\right| \leq \left|k_1 - \tilde{k_1}\right| e^{-t} + \left|k_2 - \tilde{k_2}\right| e^{-2t} \underset{t \to \infty}{\to} 0, \tag{38}$$

and the solutions are asymptotically stable, once again.

Finally, the solution for species S_3 is

$$h(t) = (2c_1 - c_2)e^{-t} + c_1 e^{-2t} + (c_1 + c_2 + c_3) = p_1 e^{-t} + p_2 e^{-2t} + p_3, \tag{39}$$

and in this case:

$$\begin{aligned} \left|h(t) - \tilde{h}(t)\right| &= \left|(p_1 e^{-t} + p_2 e^{-2t} + p_3) - (\tilde{p_1} e^{-t} + \tilde{p_2} e^{-2t} + \tilde{p_3})\right| \leq |p_1 - \tilde{p_1}| e^{-t} \\ &+ |p_2 - \tilde{p_2}| e^{-2t} + |p_3 - \tilde{p_3}| \underset{t \to \infty}{\to} |p_3 - \tilde{p_3}| \neq 0, \end{aligned} \tag{40}$$

and the solutions are stable, however, they are not asymptotically stable.

Chemically, this implies that small errors in the weights and/or dilutions tend to disappear over time in the case of the species S_1 and S_2, but not in the case of the species S_3. In general, these small experimental errors will remain bounded throughout the reaction, and they will be small if the initial error only produces a slight difference, for example, [8, 9].

8. Several inverse problems and questions about $\mathcal{F}_{FOCKM}$

The direct problem we have considered is as follows: given a certain mechanism of chemical reactions (a FOCKM in our case), we want to describe it mathematically through an ODE system and then find the solutions for certain initial conditions. The inverse problem that we are going to consider consists of giving a non-negative function $f(t)$, appropriate (i.e., $f \in \mathcal{L}$), and trying to find, if it exists, the corresponding FOCKM problem such that $f(t)$ can be one of its solutions. This is called a causation inverse problem (see, e.g., [10, 11], or [12]). Usually inverse problems, whatever their type (e.g., [11] or [12]), raise interesting questions of existence, uniqueness, and stability. In the next subsections, we analyze these topics.

8.1 Existence

The first question that arises is whether, for any function $f(t)$ which is a linear combination of exponentials belonging to $E_{\leq 0}$ and/or *EPO* functions, there exists a

FOCKM problem that has $f(t)$ as one of its solutions. This has already been answered negatively, since only non-negative functions can be solutions of the ODE system.

Then, refining the question, we can ask ourselves if functions of the form: e^{0t}, e^{-kt}, $t^p e^{-kt}$, $e^{-at}\cos(bt)$, $e^{-at}\sin(bt)$, $t^p e^{-at}\cos(bt)$, and $t^p e^{-at}\sin(bt)$, with $k > 0, p > 0$, $a > 0$ and $b \in \mathbb{R}$ always have their corresponding FOCKM problem. This new question has an affirmative answer, as it was observed in Section 5.

Finally, let us consider a certain non-negative function with given coefficients, for example $f(t) = 2e^{-t} - e^{-2t} - e^{-2t}\cos(6t) + 3$. Our last question is whether or not it is possible to obtain a FOCKM problem that gives exactly that solution, with those coefficients. In other words, the question is not to obtain functions of a certain form but exactly a given function with given coefficients. This more demanding—and therefore more complicated—inverse problem is a research challenge for further work. Considering the results of Section 5, it seems reasonable to conjecture that the answer could be affirmative.

8.2 Uniqueness

The first question that arises is whether two different FOCKMs can give solutions of the same type. In order to answer this question, it would be enough to consider Example N° 4, represented in **Figure 8**.

The corresponding ODE system is

$$\begin{cases} \dfrac{d[S_1]}{dt} = -k_{12}[S_1] + k_{21}[S_2] \\ \dfrac{d[S_2]}{dt} = k_{12}[S_1] - k_{21}[S_2] \end{cases}, \tag{41}$$

and

$$\mathbb{A} = \begin{pmatrix} -k_{12} & k_{21} \\ k_{12} & -k_{21} \end{pmatrix} \tag{42}$$

And since the eigenvalues are: $\lambda = 0$ and $\lambda = -k_{12} - k_{21}$, the solutions are of the form: $f(t) = c_1 e^{0t} + c_2 e^{-(k_{12}+k_{21})t}$. As a consequence, if $k_{12} = 4$ and $k_{21} = 1$ or if $k_{12} = 3$ and $k_{21} = 2$, in both cases the solutions are of the form $f(t) = c_1 e^{0t} + c_2 e^{-5t}$. Then, different FOCKMs can give the same eigenvalues and multiplicities, and therefore, functions of the same form.

A second, more refined question is whether the same eigenvectors can be obtained, in addition to having the same eigenvalues. For this purpose, it is interesting to consider Example N° 5, schematized in **Figure 9**.

For the FOCKM in **Figure 9a**, the following ODE system is obtained:

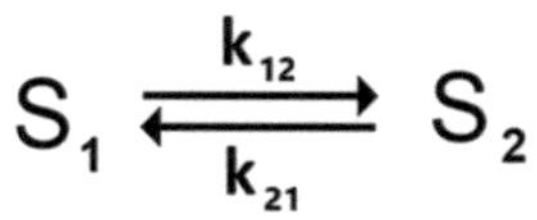

Figure 8.
FOCKM with opposite reactions.

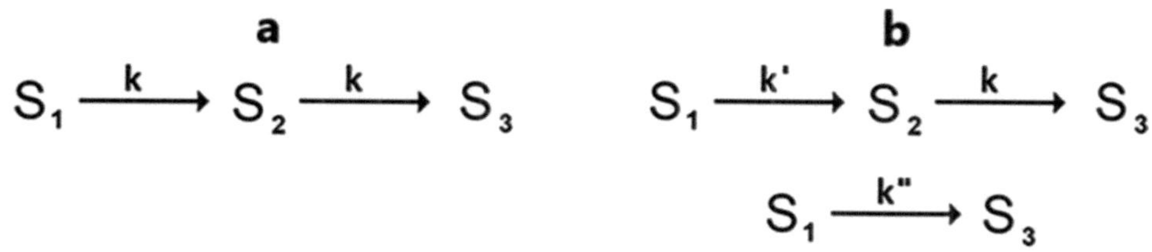

Figure 9.
Two different FOCKMs involving three chemical species.

$$\begin{cases} \dfrac{d[S_1]}{dt} = -k[S_1] \\ \dfrac{d[S_2]}{dt} = k[S_1] - k[S_2] \\ \dfrac{d[S_3]}{dt} = k[S_2] \end{cases}, \tag{43}$$

and

$$\mathbb{A} = \begin{pmatrix} -k & 0 & 0 \\ k & -k & 0 \\ 0 & k & 0 \end{pmatrix} \tag{44}$$

And the eigenvalues are: $\lambda_1 = -k$ and $\lambda_2 = 0$, being the associated eigenvectors $\overrightarrow{v_1} = (\,0 \quad -1 \quad 1\,)$ and $\overrightarrow{v_2} = (\,0 \quad 0 \quad 1\,)$.

In the case of the FOCKM in **Figure 9b**, the following ODE system is obtained:

$$\begin{cases} \dfrac{d[S_1]}{dt} = -(k' + k'')[S_1] \\ \dfrac{d[S_2]}{dt} = k'\,[S_1] - k[S_2] \\ \dfrac{d[S_3]}{dt} = k''\,[S_1] + k\,[S_2] \end{cases}, \tag{45}$$

additionally, if we consider the case $k' + k'' = k$, then the FOCKM matrix is:

$$\mathbb{A} = \begin{pmatrix} -k & 0 & 0 \\ k' & -k & 0 \\ k'' & k & 0 \end{pmatrix} \tag{46}$$

In this case, the eigenvalues are: $\lambda_1 = 0$ and $\lambda_2 = -k$, being the associated eigenvectors: $\overrightarrow{v_1} = (\,0 \quad 0 \quad 1\,)$ and $\overrightarrow{v_2} = (\,0 \quad -1 \quad 1\,)$.

Then, both FOCKM problems give the same eigenvalues and eigenvectors, with a change in the order.

Thus, we know that different FOCKMs can give the same eigenvalues and eigenvectors, but can they give exactly the same solution with the same coefficients? To analyze this question, let us consider the following example in four parts:

The FOCKM in **Figure 10a** is an example of successive reactions, and its corresponding mathematical model is

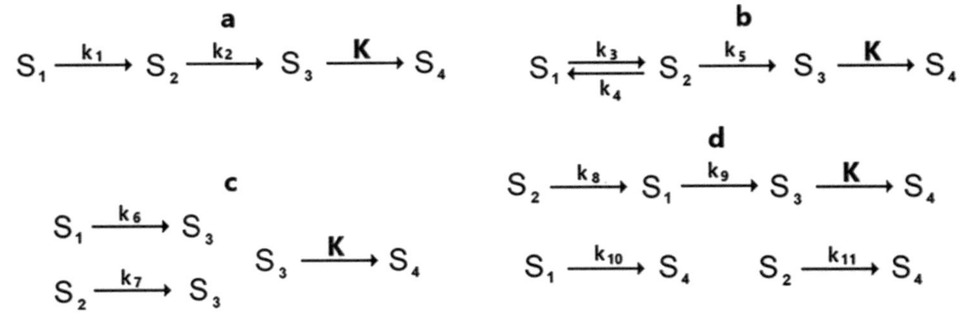

Figure 10.
Four different FOCKMs involving several chemical species.

$$\begin{cases} \dfrac{d[S_1]}{dt} = -k_1[S_1] \\ \dfrac{d[S_2]}{dt} = k_1[S_1] - k_2[S_2] \\ \dfrac{d[S_3]}{dt} = k_2[S_2] - K[S_3] \\ \dfrac{d[S_4]}{dt} = K[S_3] \end{cases}, \tag{47}$$

and

$$\mathbb{A} = \begin{pmatrix} -k_1 & 0 & 0 & 0 \\ k_1 & -k_2 & 0 & 0 \\ 0 & k_2 & -K & 0 \\ 0 & 0 & K & 0 \end{pmatrix} \tag{48}$$

If $k_1 = 4$, $k_2 = 1$, and $K = 2$, the eigenvalues are: $\lambda_1 = 0$, $\lambda_2 = -1$, $\lambda_3 = -4$ and $\lambda_4 = -2$, being the associated eigenvectors: $\vec{v_1} = (0 \quad 0 \quad 0 \quad 1)$, $\vec{v_2} = (0 \quad -1/2 \quad -1/2 \quad 1)$, $\vec{v_3} = (-3 \quad 4 \quad -2 \quad 1)$ and $\vec{v_4} = (0 \quad 0 \quad -1 \quad 1)$, respectively.

Consequently, the general solution is

$$\begin{pmatrix} [S_1](t) \\ [S_2](t) \\ [S_3](t) \\ [S_4](t) \end{pmatrix} = c_1 e^{0t} \begin{pmatrix} 0 \\ 0 \\ 0 \\ 1 \end{pmatrix} + c_2 e^{-t} \begin{pmatrix} 0 \\ -1 \\ -1 \\ 2 \end{pmatrix} + c_3 e^{-4t} \begin{pmatrix} -3 \\ 4 \\ -2 \\ 1 \end{pmatrix} + c_4 e^{-2t} \begin{pmatrix} 0 \\ 0 \\ -1 \\ 1 \end{pmatrix} \tag{49}$$

If the initial conditions are

$$[S_1](0) = 0, [S_2](0) = 0, [S_3](0) = 1, \text{and } [S_4](0) = 1, \tag{50}$$

then the solution of the FOCKM problem is:

$$[S_1](0) = 0, [S_2](0) = 0, [S_3](0) = e^{-2t}, \text{and } [S_4](0) = 2 - e^{-2t}. \tag{51}$$

Now, the FOCKM in **Figure 10b** is an example of a reversible reaction, followed by successive reactions. The corresponding mathematical model is

$$\begin{cases} \dfrac{d[S_1]}{dt} = -k_3[S_1] + k_4[S_2] \\ \dfrac{d[S_2]}{dt} = k_3[S_1] - (k_4 + k_5)[S_2] \\ \dfrac{d[S_3]}{dt} = k_5[S_2] - K[S_3] \\ \dfrac{d[S_4]}{dt} = K[S_3] \end{cases}, \tag{52}$$

and

$$\mathbb{A} = \begin{pmatrix} -k_3 & k_4 & 0 & 0 \\ k_3 & -(k_4 + k_5) & 0 & 0 \\ 0 & k_5 & -K & 0 \\ 0 & 0 & K & 0 \end{pmatrix} \tag{53}$$

If $k_3 = 8$, $k_4 = 1$, $k_5 = 5$, and $K = 2$, the eigenvalues are: $\lambda_1 = 0$, $\lambda_2 = -2$, $\lambda_3 = -10$ and $\lambda_4 = -4$, being the associated eigenvectors: $\overrightarrow{v_1} = (\,0 \quad 0 \quad 0 \quad 1\,)$, $\overrightarrow{v_2} = (\,0 \quad 0 \quad -1 \quad 1\,)$, $\overrightarrow{v_3} = (\,-4 \quad 8 \quad -5 \quad 1\,)$ and $\overrightarrow{v_4} = (\,1/5 \quad 4/5 \quad -2 \quad 1\,)$, respectively.

Then, the general solution is

$$\begin{pmatrix} [S_1](t) \\ [S_2](t) \\ [S_3](t) \\ [S_4](t) \end{pmatrix} = c_1 e^{0t} \begin{pmatrix} 0 \\ 0 \\ 0 \\ 1 \end{pmatrix} + c_2 e^{-2t} \begin{pmatrix} 0 \\ 0 \\ -1 \\ 1 \end{pmatrix} + c_3 e^{-10t} \begin{pmatrix} -4 \\ 8 \\ -5 \\ 1 \end{pmatrix} + c_4 e^{-4t} \begin{pmatrix} 1 \\ 4 \\ -10 \\ 5 \end{pmatrix} \tag{54}$$

If the initial conditions are

$$[S_1](0) = 0, [S_2](0) = 0, [S_3](0) = 1, \text{and } [S_4](0) = 1, \tag{55}$$

then the solution of the FOCKM problem is:

$$[S_1](0) = 0, [S_2](0) = 0, [S_3](0) = e^{-2t}, \text{and } [S_4](0) = 2 - e^{-2t}. \tag{56}$$

We consider now the FOCKM in **Figure 10c**, which consists of competitive reactions plus one more FOCR. The corresponding mathematical model is

$$\begin{cases} \dfrac{d[S_1]}{dt} = -k_6[S_1] \\ \dfrac{d[S_2]}{dt} = -k_7[S_2] \\ \dfrac{d[S_3]}{dt} = k_6[S_1] + k_7[S_2] - K[S_3] \\ \dfrac{d[S_4]}{dt} = K[S_3] \end{cases}, \tag{57}$$

and

$$\mathbb{A} = \begin{pmatrix} -k_6 & 0 & 0 & 0 \\ 0 & -k_7 & 0 & 0 \\ k_6 & k_7 & -K & 0 \\ 0 & 0 & K & 0 \end{pmatrix} \tag{58}$$

If $k_6 = 4$, $k_7 = 6$, and $K = 2$, the eigenvalues are: $\lambda_1 = 0$, $\lambda_2 = -2$, $\lambda_3 = -6$ and $\lambda_4 = -4$, being the associated eigenvectors: $\overrightarrow{v_1} = (0 \quad 0 \quad 0 \quad 1)$, $\overrightarrow{v_2} = (0 \quad 0 \quad -1 \quad 1)$, $\overrightarrow{v_3} = (0 \quad 2 \quad -3 \quad 1)$ and $\overrightarrow{v_4} = (1 \quad 0 \quad -2 \quad 1)$, respectively. As a consequence, the general solution is

$$\begin{pmatrix} [S_1](t) \\ [S_2](t) \\ [S_3](t) \\ [S_4](t) \end{pmatrix} = c_1 e^{0t} \begin{pmatrix} 0 \\ 0 \\ 0 \\ 1 \end{pmatrix} + c_2 e^{-2t} \begin{pmatrix} 0 \\ 0 \\ -1 \\ 1 \end{pmatrix} + c_3 e^{-6t} \begin{pmatrix} 0 \\ 2 \\ -3 \\ 1 \end{pmatrix} + c_4 e^{-4t} \begin{pmatrix} 1 \\ 0 \\ -2 \\ 1 \end{pmatrix} \tag{59}$$

If the initial conditions are:

$$[S_1](0) = 0, [S_2](0) = 0, [S_3](0) = 1, \text{and } [S_4](0) = 1, \tag{60}$$

then the solution of the FOCKM problem is:

$$[S_1](0) = 0, [S_2](0) = 0, [S_3](0) = e^{-2t}, \text{and } [S_4](0) = 2 - e^{-2t}. \tag{61}$$

Finally, the FOCKM in **Figure 10d** consists of successive reactions and two parallel reactions. The corresponding mathematical model is

$$\begin{cases} \dfrac{d[S_1]}{dt} = -(k_9 + k_{10})[S_1] + k_8[S_2] \\ \dfrac{d[S_2]}{dt} = -(k_8 + k_{11})[S_2] \\ \dfrac{d[S_3]}{dt} = k_9[S_1] - K[S_3] \\ \dfrac{d[S_4]}{dt} = k_{10}[S_1] + k_{11}[S_2] + K[S_3] \end{cases}, \tag{62}$$

and

$$\mathbb{A} = \begin{pmatrix} -(k_9 + k_{10}) & k_8 & 0 & 0 \\ 0 & -(k_8 + k_{11}) & 0 & 0 \\ k_9 & 0 & -K & 0 \\ k_{10} & k_{11} & K & 0 \end{pmatrix} \tag{63}$$

If $k_8 = 1$, $k_9 = 1$, $k_{10} = 3$, $k_{11} = 2$, and $K = 2$, the eigenvalues are: $\lambda_1 = 0$, $\lambda_2 = -2$, $\lambda_3 = -4$ and $\lambda_4 = -3$, being the associated eigenvectors: $\overrightarrow{v_1} = (0 \quad 0 \quad 0 \quad 1)$, $\overrightarrow{v_2} = (0 \quad 0 \quad -1 \quad 1)$, $\overrightarrow{v_3} = (-2 \quad 0 \quad 1 \quad 1)$ and $\overrightarrow{v_4} = (-1 \quad -1 \quad 1 \quad 1)$, respectively.

So, the general solution is

$$\begin{pmatrix} [S_1](t) \\ [S_2](t) \\ [S_3](t) \\ [S_4](t) \end{pmatrix} = c_1 e^{0t} \begin{pmatrix} 0 \\ 0 \\ 0 \\ 1 \end{pmatrix} + c_2 e^{-2t} \begin{pmatrix} 0 \\ 0 \\ -1 \\ 1 \end{pmatrix} + c_3 e^{-4t} \begin{pmatrix} -2 \\ 0 \\ 1 \\ 1 \end{pmatrix} + c_4 e^{-3t} \begin{pmatrix} -1 \\ -1 \\ 1 \\ 1 \end{pmatrix} \quad (64)$$

If the initial conditions are

$$[S_1](0) = 0, [S_2](0) = 0, [S_3](0) = 1, \text{and } [S_4](0) = 1, \quad (65)$$

then the solution of the FOCKM problem is:

$$[S_1](0) = 0, [S_2](0) = 0, [S_3](0) = e^{-2t}, \text{and } [S_4](0) = 2 - e^{-2t}. \quad (66)$$

As it can be observed, all these FOCKMs are mathematically different since they have different eigenvalues and eigenvectors and come from different FOCKM matrices. They are also chemically different, since they correspond to successive reactions, opposite and successive reactions, competitive reactions followed by another FOCR, and successive reactions with parallel reactions. However, given the same initial conditions, they all give the exact same solution.

Then, it can be concluded that the question of uniqueness has a negative answer. However, before reaching a definitive conclusion, it should be noted that in these FOCKM problems, the chemical species S_1 and S_2 do not appear at the end because $[S_1](0) = 0$, $[S_2](0) = 0$, $[S_3](0) = 1$, and $[S_4](0) = 1$ (50–55–60–65), i.e., the only reaction that takes place in all cases is that one in **Figure 11**.

In other words, given the initial conditions considered (50–55–60–65), the equality of the results is chemically trivial.

Is it possible to find examples of non-uniqueness that are not chemically trivial? So far, we do not have an answer, but we can conjecture that the answer to that question would be negative. Indeed, it seems logical to assume that if there are non-trivial chemical differences, the solutions would be different.

8.3 Stability

The last issue to consider is the stability of the inverse problem, that is, if two solutions differ in norm by less than an appropriate δ, i.e., $\|f - g\|_{\infty,[0,+\infty)} < \delta$, then it is possible to affirm that they come from FOCKM matrices $\mathbb{A}$, $\mathbb{B}$ whose difference is less than epsilon in some matrix norm, i.e., $\|\mathbb{A} - \mathbb{B}\|_M$ (being $\|\bullet\|_M$ some appropriate norm of matrices).

Then, if we consider chemically trivial problems like those in **Figure 10a–d**, the solutions are the same, and so $\|f - g\|_{\infty,[0,+\infty)} = 0 < \delta$, for every δ, positive, and since they come from different FOCKM matrices (48–53–58–63) and therefore, $\|\mathbb{A} - \mathbb{B}\|_M$ will not be less than ε, if a suitable value of ε is chosen.

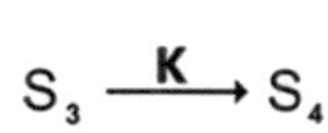

Figure 11.
FOCR involving two chemical species.

The question that remains is whether stability conditions can be obtained if chemically trivial situations are avoided. In this case it is much more difficult to establish a conjecture intuitively. Once again, this is a problem for further research. In relation to the above, a practical question that arises is the following: If two solutions are slightly different, that is, $\|f - g\|_{\infty,[0,+\infty)} < \delta$, for a small $\delta > 0$, is there an experimental procedure that allows us to decide between two possible mechanisms?

Let us illustrate how to proceed, returning to the FOCKMs in **Figure 9**, giving values to $k' = k'' = 1$, and $k = k' + k'' = 2$.

So, for the case 9a we have

$$\begin{cases} \dfrac{d[S_1]}{dt} = -2[S_1] \\ \dfrac{d[S_2]}{dt} = 2[S_1] - 2[S_2] \cdot \\ \dfrac{d[S_3]}{dt} = 2[S_2] \end{cases} \tag{67}$$

Thus, solving the system, the result is

$$[S_1](t) = x_0 e^{-2t}, [S_2](t) = y_0 e^{-2t} + 2x_0 t e^{-2t}, \text{and } [S_3](t) = (x_0 + y_0 + z_0) - (x_0 + y_0)e^{-2t} - 2x_0 t e^{-2t}. \tag{68}$$

Now, for the case 9b we have

$$\begin{cases} \dfrac{d[S_1]}{dt} = -2[S_1] \\ \dfrac{d[S_2]}{dt} = [S_1] - 2[S_2] \cdot \\ \dfrac{d[S_3]}{dt} = [S_1] + 2[S_2] \end{cases} \tag{69}$$

Solving the system, the result is:

$$[S_1](t) = x_0 e^{-2t}, [S_2](t) = y_0 e^{-2t} + x_0 t e^{-2t}, \text{and } [S_3](t) = (x_0 + y_0 + z_0) - (x_0 + y_0)e^{-2t} - x_0 t e^{-2t}. \tag{70}$$

Then, for general initial conditions: $[S_1](0) = x_0$, $[S_2](0) = y_0$, $[S_3](0) = z_0$, there is no difference in the concentration of S_1, in the concentration of S_2 the difference is $\varphi(t) = x_0 t e^{-2t}$, and in the concentration of species S_3 the difference is again $\varphi(t) = x_0 t e^{-2t}$. It should be noted that the maximum of $\varphi(t)$ is obtained at $t = 1/2$ and its value is $\varphi(1/2) = \frac{x_0}{2e}$. For that reason, the strategy to discern which of the two mechanisms is the correct, consists of starting from the highest possible initial concentration of S_1 and measuring the concentrations of the three species at time $t = 1/2$, when the greatest differences can be observed.

9. Conclusion

In this work, the set of functions $\mathcal{F}_{FOCKM}$ which consists in all possible theoretical solutions of first-order chemical kinetics problems has been studied. It was seen that it was impossible to have a vector space structure; however, it is possible to define an appropriate norm for that set.

Theoretical problems related to $\mathcal{F}_{FOCKM}$ have been studied, as well as practical issues related to the propagation of errors in chemical experiments.

Finally, several inverse problems related to this set of functions are included, and an interesting conjecture is provided.

Additionally, some open questions remain pending, for instance:

1. Let $\Lambda = \{\lambda_1, \lambda_2, \cdots, \lambda_n\}$ be a set of complex numbers, such that $Re(\lambda_i) \leq 0, \forall i = 1, 2, \ldots, n$. Is it possible to propose a FOCKM whose eigenvalues are exactly $\lambda_1, \lambda_2, \cdots, \lambda_n$?

2. If the first question has an affirmative answer, let us consider a set of algebraic multiplicities $\mathrm{A} = \{a_1, a_2, \cdots, a_n\}$ and another set of geometric multiplicities $\mathrm{G} = \{g_1, g_2, \cdots, g_n\}$, being $g_i \leq a_i, \forall i = 1, 2, \ldots, n$. Is it possible to propose a FOCKM such that the eigenvalues algebraic multiplicities are $a_1, a_2, \cdots, a_n$ and the geometric ones are $g_1, g_2, \cdots, g_n$?

These open questions will be the subject of further research to be presented in future works.

Acknowledgements

The author wishes to thank the anonymous reviewer for his/her helpful suggestions for improving the chapter and also thank Marjorie Chaves for reviewing this work.

Author details

Victor Martinez-Luaces
University of the Republic of Uruguay, Montevideo, Uruguay

*Address all correspondence to: victorml@fing.edu.uy

References

[1] Martinez-Luaces V. Eigenvalues of matrices in chemical kinetics and their algebraic and geometric multiplicities [Internet]. In: Nonlinear Systems and Matrix Analysis - Recent Advances in Theory and Applications. London, UK: IntechOpen; 2024. DOI: 10.5772/intechopen.1003778

[2] Martinez-Luaces V. Concentration curves in first-order chemical kinetics: All possible cases. In: A Closer Look at Chemical Kinetics. New York: Nova Science Publishers; 2023. pp. 13-28

[3] Roman S. Advanced Linear Algebra. Berlin Heidelberg: Springer-Verlag; 2013

[4] Ricardo HJ. A Modern Introduction to Differential Equations. Amsterdam: Elsevier Science; 2020

[5] Zaitsev VF, Polyanin AD. Handbook of Exact Solutions for Ordinary Differential Equations. Boca Raton, FL: CRC Press; 2002

[6] Varga RS. Geršgorin and his Circles. Berlin Heidelberg: Springer-Verlag; 2011

[7] Michel AN, Hou L, Liu D. Stability of Dynamical Systems: Continuous, Discontinuous, and Discrete Systems. Boston: Springer; 2008

[8] Martinez-Luaces V. Stability of ODE systems associated with first-order chemical kinetics mechanisms with and without final products. Konuralp Journal of Mathematics. 2016;**4**(1):80-87

[9] Martinez-Luaces V. Matrices in chemical problems: Characterization, properties and consequences about the stability of ODE Systems. In: Baswell AR, editor. Advances in Mathematics Research. New York: Nova Science Publishers; 2017. pp. 1-33

[10] Martinez-Luaces V, Fernández-Plaza JA, Rico L. About the notion of inverse problem in STEM education. In: Active Learning - Research and Practice for STEAM and Social Sciences Education. IntechOpen; 2022. DOI: 10.5772/intechopen.106479

[11] Groetsch CW. Inverse Problems: Activities for Undergraduates. UK: Cambridge University Press; 1999

[12] Groetsch CW. Inverse problems: The other two-thirds of the story. Quaestiones Mathematicae. 2001;**24**(1): 89-94

Chapter 4

Transition from Maturity to Risk-Based Cybersecurity: A Cyber-Process Hazard Assessment Methodology

Victor Siang Tze Sim, Kok Hwa Lim and Tan Yong Rui

Abstract

Industrial control systems (ICS) face unprecedented cybersecurity challenges as Industry 4.0 technologies blur traditional boundaries between safety and cybersecurity. Current maturity-based cybersecurity approaches create unsustainable overhead, while traditional process hazard analysis (PHA) methodologies fail to account for deliberate cyberattacks. This chapter addresses the critical need for integrated safety-security risk assessment frameworks that enable risk-based prioritization of critical assets. We propose a modified cyber-process hazard assessment (cyber-PHA) methodology integrating layer of protection analysis (LOPA) with cyber risk quantification using the factor analysis of information risk (FAIR) framework. The approach was applied on a pilot-scale water treatment testbed using an example of scenario that involved excess oxidant damaging membrane barriers through cyber-induced failures. Semi-quantitative vulnerability assessments were conducted using ISA 62443 security levels (SL-0 to SL-4) combined with exploit prediction scoring system (EPSS) data for threat event frequency estimation. The integrated cyber-PHA methodology enables transition from broad maturity-based security models to targeted, risk-based frameworks. This approach provides practitioners with familiar LOPA-based tools while incorporating objective cyber risk quantification, supporting economical protection strategies until comprehensive operational technology (OT) domain statistical data becomes available.

Keywords: cyber-physical risk assessment, process hazard analysis, risk-based security, cyber risk quantification, FAIR methodology

1. Introduction

The digital transformation of industrial processes through Industry 4.0 technologies has fundamentally altered the cybersecurity landscape for industrial control systems (ICS), creating unprecedented challenges at the intersection of safety and security. Traditional safety and security practitioners have historically operated within separate domains, each employing distinct frameworks, terminologies, and methodologies. However, as ICS become increasingly interconnected, this separation has become not only impractical but potentially dangerous, as cyberattacks can now be possible to cause safety incidents directly [1].

A fundamental controversy exists in how organizations should prioritize cybersecurity investments in safety-critical systems. The maturity-based cybersecurity paradigm advocates for building the highest level of defense around all assets [2, 3], following standards like ISA 62443 to achieve uniform security levels. However, this approach conflicts with risk-based methodologies that emphasize targeted protection of critical assets based on their potential impact [2]. The ISA-TR84.00.09-2023 technical report acknowledges this tension by proposing cyber-process hazard assessment (cyber-PHA) methodologies, yet these approaches have been criticized as overly conservative, assuming immediate success of cyber-attacks and complete failure of independent protection layers (IPLs) without considering existing cybersecurity measures.

Another significant divergence lies in the quantification of cyber risks within operational technology (OT) environments. Traditional IT-focused frameworks like the factor analysis of information risk (FAIR) methodology offer structured approaches to cyber risk quantification, but their applicability to OT systems remains contentious due to the lack of statistical data on cyberattacks in process industries. This creates a methodological divide between those advocating for semi-quantitative approximations and practitioners demanding empirically based risk calculations.

This chapter addresses the critical need for integrated safety-security risk assessment frameworks that enable economical, risk-based prioritization of critical assets while maintaining compatibility with established safety lifecycle practices. The work demonstrates how cyber risk quantification can be systematically integrated with traditional layer of protection analysis (LOPA) methodologies, providing practitioners with familiar tools while incorporating objective cyber risk assessment techniques.

The significance of this work extends beyond academic discourse to practical industrial application. The proposed methodology offers a bridge between the established safety engineering community and emerging cybersecurity requirements, potentially transforming how organizations approach risk management in digitally integrated industrial systems.

The main aim of this work is to develop and demonstrate a modified cyber-process hazard assessment methodology that integrates LOPA with cyber risk quantification [4] using the FAIR framework, enabling the transition from broad maturity-based security models to targeted, risk-based frameworks. The approach is validated through application to a pilot-scale water treatment testbed, examining scenarios involving excess oxidant damaging membrane barriers through cyber-induced failures.

The principal conclusions demonstrate that semi-quantitative vulnerability assessments using ISA 62443 Security Levels combined with dynamic threat intelligence from the exploit prediction scoring system (EPSS) can provide more accurate and economical risk assessments than conservative cyber-PHA approaches. The integrated methodology enables practitioners to move beyond the assumption of complete system compromise to nuanced risk evaluations that account for actual cybersecurity controls and threat landscapes. This approach provides a practical pathway for organizations to implement risk-based cybersecurity strategies while maintaining alignment with established safety engineering practices, supporting economical protection strategies until comprehensive OT domain statistical data becomes available.

2. Cyber-PHA risk assessment

PHA have not considered cyber-attacks to be possible due to the nature of the initiating events. For safety, it is mainly due to failure of equipment, human actions, etc., while for cybersecurity, it is a deliberate attempt to compromise the system. As such, for cybersecurity, it is usually studied under security and vulnerability assessment. The two communities worked separately with their own frameworks and terminologies. With the digital revolution, safety and security risks have to be prioritized concurrently as they are more intertwined than before. One of the ways is to perform a top-down approach through the identification of services and functions to protect [1, 5].

ISA-TR84.00.09-2023 Part 1 is a technical report [4] that provides guidance on how to implement cybersecurity within the IEC-61511 [6] and ISA-84.00.01-2004 lifecycle. IEC standard 61511 and ISA-84.00.01-2004 are technical standards that establish engineering practices designed to ensure the safe and reliable operation of industrial processes that involve the use of instrumentation. Cyber-PHA risk assessment methodology utilizes a hazard and operability study (HAZOP) approach, but it evaluates the cyber nodes that represent the cyber assets in the zones and conduits design. While concern is on the process control, alarms, and interlocks (PCAI), the entire operational technology (OT) plant network including wireless access points, should be addressed as they can be pathways to compromise the ICS. In summary, these are the steps [4]:

- Discard those initiating events (IE) that are not susceptible to cyber influence and review each IE. If it can be caused by a cyber-attack, a high demand of one event per year is assumed for convenience. If it cannot be caused by a cyber-attack, IE frequency is 0 per year.
- For the IE, review the independent protection layers (IPLs) to identify safeguards susceptible to cyber influence and the corresponding probability of failure upon demand (PFD) = 1.
- Recalculate hazard frequency and perform LOPA calculations with applicable IE frequency and safeguard PFDs.

It is clear this approach can be highly conservative as it assumes the immediate success of the cyber-attack and failure of the IPL upon demand without consideration of any cybersecurity measures in place.

In industries that use IEC standard 61511 and ISA-84.00.01-2004, the ISA 62443 series of standards is commonly applied to address cybersecurity in ICS. ISA 62443-3-2 [7] defines the risk-based security levels (SLs) that are used to set appropriate technical security capabilities. ISA-62443-4-2 [8] defines four different levels (1, 2, 3, and 4) of SLs, each with an increasing level of security.

- SL 0: Assumes no need for security measures or protection mechanisms.
- SL 1: Provides defense against accidental or opportunistic breaches.
- SL 2: Guards against deliberate breaches carried out with simple tools and skills, limited resources, and low motivation.

- SL 3: Protects against targeted attacks using advanced tools and specialized ICS skills, moderate resources with moderate motivation.
- SL 4: Protects against highly sophisticated and targeted attacks using highly sophisticated tools and deep ICS skills, significant resources, and strong motivation.

ISA-62443-3-3 [9] further categorized SLs into target, achieved, and capability.

- SL-targets (SL-Ts) represent the intended security levels for a specific ICS, zone, or conduit, determined through a risk assessment of the System Under Consideration (SUC).
- Achieved SLs (SL-As) refer to the actual security levels attained by a system, assessed after its design or implementation.
- SL-capabilities (SL-Cs) indicate the security levels that system components within the SUC can inherently support when properly configured and integrated, without relying on additional compensating controls.

System requirements (SRs) and requirement enhancements (REs) function as a checklist to meet the SLs after the SUCs are arranged in terms of zones and conduits. As per ISA 62443-1-1 [10], a total of seven foundation requirements (FRs) is achieved through a series of SRs and REs:

- FR-1: Identification and authentication control (IAC),
- FR-2: Use control (UC),
- FR-3: System integrity (SI),
- FR-4: Data confidentiality (DC),
- FR-5: Restricted data flow (RDF),
- FR-6: Timely response to events (TRE), and
- FR-7: Resource availability (RA).

Using FR-4 as an illustration, there are three SRs defined and achieved through specified REs if necessary to achieve the necessary SLs.

- SR-4.1: Information confidentiality—The control system must support the ability to safeguard the confidentiality of information that requires explicit read authorization, whether the data is stored or transmitted.
    - RE-1: Ensure confidentiality for data at rest or during transmission over untrusted networks
    - RE-2: Maintain confidentiality when data crosses zone boundaries

SRs and REs	SL 1	SL 2	SL 3	SL 4
SR-4.1	✓	✓	✓	✓
RE-1		✓	✓	✓
RE-2				✓
SR-4.2		✓	✓	✓
RE-1			✓	✓
SR-4.3	✓	✓	✓	✓

Table 1.
SEs and REs required for SL defined by ISA 62443-3-3 [9].

- SR-4.2: Information persistence—The control system must allow for the secure removal of all information requiring explicit read authorization from any components that are being decommissioned or taken out of active use.

  o RE-1: Secure deletion of data from shared memory resources

- SR-4.3: Use of cryptography—The control system must implement cryptographic algorithms, key lengths, and key exchange mechanisms in accordance with widely accepted security industry standards and best practices.

Table 1 shows the SRs and REs required to achieve the relevant SL according to ISA 62443-3-3. For example, SR-4.1 and SR-4.3 must be fulfilled for SL-1, and additionally, SR-4.1 (RE-1) and SR-4.2 must be fulfilled for SL-2.

3. Layer of protection analysis (LOPA)

The conduct of safety risk assessments in ICS is paramount to ensure operational reliability and safety. Traditionally, these assessments have concentrated on physical failures. However, with the increasing integration of digital technologies, cyber-induced failures have become a significant concern [11].

To systematically identify potential hazards in an industrial setting, a PHA such as a HAZOP is typically conducted. While PHA effectively identifies hazards, it does not assess whether existing safeguards are sufficient. This gap is addressed by layer of protection analysis (LOPA), which builds upon qualitative studies like HAZOP. LOPA evaluates whether additional layers of protection, such as safety instrumented systems (SIS), are required to achieve acceptable risk levels. Independent protection layers (IPLs) are used to quantify risk reduction for specific hazardous scenarios. As a simplified form of numerical risk assessment, LOPA provides an estimate of the risk associated with straightforward scenarios [12], helping determine if the existing safeguards are adequate or if further measures are necessary.

To perform LOPA analysis, it is important to illustrate with a case study that can be adopted for cyber risk analysis. iTrust Secure Water Treatment (SWaT) system [13] is an ideal example as the pilot-scale testbed has been studied by multiple authors for cyber-related simulations [1, 14, 15]. HAZOP is performed on the SWaT system (**Figures 1–3**), which supplies treated drinking water to the downstream water distribution (WADI) system. This is a comprehensive and realistic model of

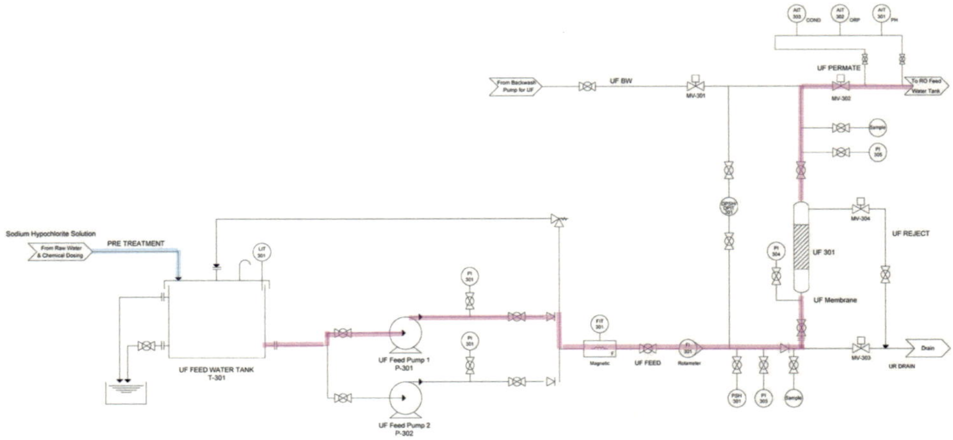

Figure 1.
Piping and instrumentation diagram (P&ID) for treatment stage P3 (ultrafiltration).

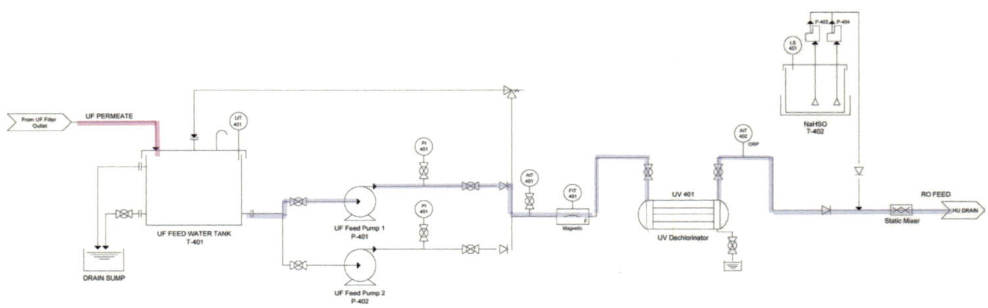

Figure 2.
Piping and instrumentation diagram (P&ID) for treatment stage P4 (dechlorination).

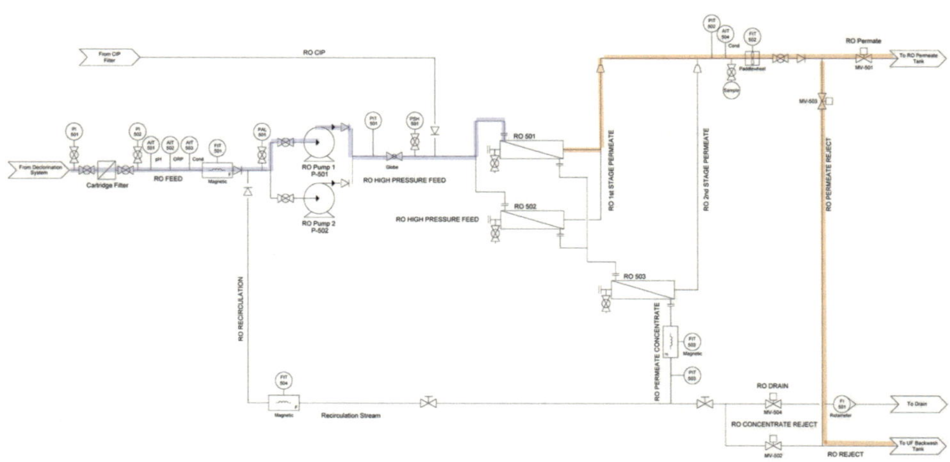

Figure 3.
Piping and instrumentation diagram (P&ID) for treatment stage P5 (RO).

DOI: http://dx.doi.org/10.5772/intechopen.1011612

Node	Element	P&ID title	Property	Deviation from intent	Process Operation mode	Cause of deviation	Effect/consequences	Initial safeguards/ mitigation
Sodium hypochlorite storage and dosing	T203 to feed water	P&ID for P1 (raw water) and P2 (chemical dosing)	Instrumentation	More than	Normal operation	Drifting of calibration of pump	Excess oxidant and overwhelm UV chlorinator. Potential damage to RO. Lead to RO shutdown, plant stoppage	1. AIT203 (ORP) High 2. AIT302 (ORP) High 3. UV401 On 4. AIT402 (ORP) High and $NaHSO_3$ dosing system 5. AIT502 (ORP) High 6. FIT502 High 7. AIT504 (Conductivity) High 8. MV501 Open and MV503 Close 9. Regular maintenance check—mitigation

Table 2.
Specific node in SWaT risk register.

a water treatment, storage, and distribution network. iTrust, a multidisciplinary research center at the Singapore University of Technology and Design (SUTD), was jointly established in 2012 by SUTD and the Singapore Ministry of Defence.

The SWaT system is a six-stage water purification facility designed to produce five gallons of potable water per minute using methods such as ultrafiltration, dechlorination, and reverse osmosis [1]. It is equipped with 68 sensors and actuators. The sensors monitor variables like tank levels, pressure, and flow rates throughout the system, while the actuators consist mainly of pumps and motorized valves. Sensor outputs are generally continuous values, whereas actuators operate in discrete states (e.g., on/off).

SWaT serves as the sole provider of potable water to the WADI system. A cyberattack that disrupts SWaT operations would impair WADI's ability to adequately store and distribute drinking water. Any functional disruption in SWaT compromises WADI's capacity to meet national standards for potable water supply, specifically those outlined in the Singapore Environmental Public Health (Water Suitable for Drinking) (No. 2) Regulations 2019 [16].

Table 2 is a risk scenario of excess oxidant damaging the final membrane barrier and a cause of potential loss of containment risk scenario for the LOPA [1]. The authors note that LOPA does possess limitations as it is not suitable to analyze scenarios with common cause failure or scenarios with multiple initiating events [12]. However, LOPA is sufficient in the evaluation of the current simple scenario used in this chapter, and the risk estimated is more conservative as it is an order-of-magnitude risk calculation with a cumulative effect on the final event frequency. For more complex scenarios involving multiple scenarios, more detailed risk analysis such as quantitative risk analysis (QRA) would be more appropriate.

3.1 Case study scenario

For this case study, a hardwired interlock design is used for alarms, and it is not subject to cyber risk. In industry, such alarms used in hazardous areas are also naturally designed to be intrinsically safe. The tolerable risk frequency is set to a low of $1 \times 10e^{-4}$ for the purpose of this study, which is referenced as the tolerable risk criteria with significant loss of containment as low as reasonably possible (ALARP) on the national level [17].

The 1993 Milwaukee Cryptosporidiosis outbreak [18] was further used as a reference by assuming a hypothetical contaminated water quality caused by deteriorated RO membranes that allow increased transmissibility of deadly pathogenic protozoa, viruses, and bacteria such as Cryptosporidium and Giardia that may result in 1–2 fatalities. The 1993 Milwaukee cryptosporidiosis outbreak marked the most extensive waterborne disease outbreak ever recorded in the history of the United States, which affected approximately 403,000 residents and resulted in at least 69 fatalities. It is believed that the contamination was due to an ineffective filtration process.

Table 3 shows the LOPA risk calculation for various scenarios. Under normal scenarios without a cyberattack, the LOPA calculations show the current IPLs are sufficient, and no further risk reduction is required. However, this situation changes in the presence of cyberattacks, and some of the IPLs are hackable, so the current risk will require a 100-risk reduction factor. This means the current cyber-PHA assessment is too conservative, that is, changes from meeting regulation requirements to not meeting them. However, this did not consider any cybersecurity measures in place, and the PFD for the IPLs can be refined for better accuracy.

Safeguards		Independent protection layers (IPLs)	Mitigated event frequency (no cyber-attack)[1]	Mitigated event frequency (cyber-attack)
Administrative control	AIT203 (ORP) High	Alarm 1—AIT203	0.2	0.2
	AIT302 (ORP) High	Alarm 2—AIT302	0.2	0.2
Elimination control	UV401 On	Dechlorinator 1 (UV401)	0.1	1
	AIT402 (ORP) High and $NaHSO_3$ dosing system	Dechlorinator 2 (*AIT402 and $NaHSO_3$ dosing system*)	0.1 × 0.1	1
	AIT502 (ORP) High	Interlock 3—AIT502 (RO shutdown)	0.1	1
Engineering control	FIT502 High	Alarm 3—FIT502[2]	0.5	1
	AIT504 (Conductivity) High	Out-of-spec water (*input bypass*)[3]	0.1 × 0.1	1
	MV501 Open and MV503 Close			
Total frequency[4]			$4 \times 10e^{-6}$	$4 \times 10e^{-2}$

[1]*For low level alarms (administrative control), 20% chance of operator to intervene. For high level alarms (elimination and engineering controls), 50% chance of operator to intervene. Mechanical or electrical failure set at 10% failure rate.*
[2]*Alarm if 20% lower with no change in feed pressure but can also set alarm for high flow which is useful as an alarm in this situation due to damaged RO membrane leading to reduced transmembrane pressure and elevated product of higher undesired conductivity.*
[3]*AIT-504 is interlocked with MV501/MV503. When AIT-504 is high due to elevated conductivity, MV501 will close and MV503 will open to divert water into drain instead of recycling back to the system.*
[4]*IPLs have to be independent. Devices linked to the same PLC will not be double counted and use the lower of the mitigated risk frequency.*

Table 3.
LOPA risk calculation in the presence of a cyberattack. Grayed-out safeguards are not included as part of LOPA calculations.

Cyber-PHA needs to consider how SLs are categorized through ISA 62443-3-2 [7]. The first step is to look at how the SWaT network architecture can be arranged into relevant zones and conduits and the respective allocation of SLs. Zones in different colored shaded zones are a collection of logical or physical assets with shared security requirements, determined by factors like criticality and potential impact. Conduits refer to sets of communication-related assets that also adhere to common security requirements. Additionally, conduits can represent communication pathways or tunnels that connect different zones [7]. **Figure 4** shows the logical separation of the SWAT assets into appropriate zones and conduits with PLC 4 and 5 indicated within purple dotted lines for improved cyber-PHA risk assessment.

4. Improved cyber-PHA risk assessment

As seen in **Table 3**—LOPA Risk Calculation, the important controls in terms of elimination and engineering controls, the sensors, and final elements are located in PLCs 4 and 5. AIT3XX or AIT4XX has the initial number that corresponds to the PLC it is connected with; that is, AIT302 is connected to PLC 3 and AIT402 is connected to

PLC 4. Assuming the PFD = 1 upon a cyber-attack makes sense if the entire network is leveled without any security considerations, that is, SL = 0 for all zones, but this is not practical. It is common for most large organizations to look at cybersecurity measures in some form, and the ISA-62443 series of standards is well-known in the process industries. A combination of SLs should be available, and the PFD should be a sequential risk aggregation of these SLs. In this respect, the SL level could be used in a semi-quantitative manner.

The level 1 devices compromised could be identified as having a higher SL. This will help to reduce the risk mitigation requirement to bring the entire risk down to an acceptable frequency. However, the attribution of PFDs to SLs is unprecedented, and statistical data for such failures due to cyberattacks are not easily available, especially in the operational technology (OT) environment. The authors propose a recommendation to propose protection of critical control at SL 4 with 0.1 PFD, which is similar to that of human error. It is assumed that the conservative approach is for the operator failing to manually intervene based on the alarm signals as a mitigation measure caused by a failure of automation due to cyber-interference. Note that the earlier hardwire interlock design was used for alarms and is non-hackable. Purple dotted lines within **Figure 4** indicate PLC 4 and 5 for improved cyber-PHA risk assessment.

Assuming the IPLs provide only interlocks, within the categories of elimination and engineering controls in **Table 2**, that is, AIT402 (ORP) High and $NaHSO_3$ dosing system, and AIT504 (Conductivity) High, and MV501 Open and MV503 Close, are the only safeguards allocated with SL 4 due to cost considerations, the eventual mitigated event frequency in the event of a cyber-attack affecting applicable IPLs will be $4 \times 10e^{-4}$. While this arguably can meet the tolerable risk frequency of $1 \times 10e^{-4}$, it will not be possible if the tolerable risk is set to a more conservative number, that is, a greater number of fatalities.

In such events, these can be mitigated by

a. Adding more non-digital IPLs equivalent or converting existing IPLs to be non-hackable, that is, hardwired alarms and interlocks, or mechanical design; or

b. Accurately quantify the cyber risk of controls. Cyber risk quantification methodologies suffer from the lack of statistical data for cyberattacks in the OT domain. A semi-quantitative approximation is introduced using the factor analysis of information risk (FAIR).

The authors note that the cost of allocating the relevant safeguards with SL 4 is unlikely to be an ALARP situation. The risk level should be proportionate to the time, trouble, cost, and physical difficulty involved, and the direct allocation to SL4 without considering the risk to be quantified can be considered conservative.

The above example illustrates the improvement cyber-PHA can pursue in a form that can adequately address IEC 61511 requirements. The availability of statistical data will enable the quantifiable risk due to cyberattacks to be more robust, and cyber risk quantification can mature toward a prominent methodology in the cyber-insurance industry. Like the growing role of climate change modeling in quantifying flood risk and informing insurance premium assessments, the availability of statistical data on cyberattacks within the process industries will be essential for accurately evaluating individual safety risks, environmental impacts, and financial exposures arising from critical business interruptions.

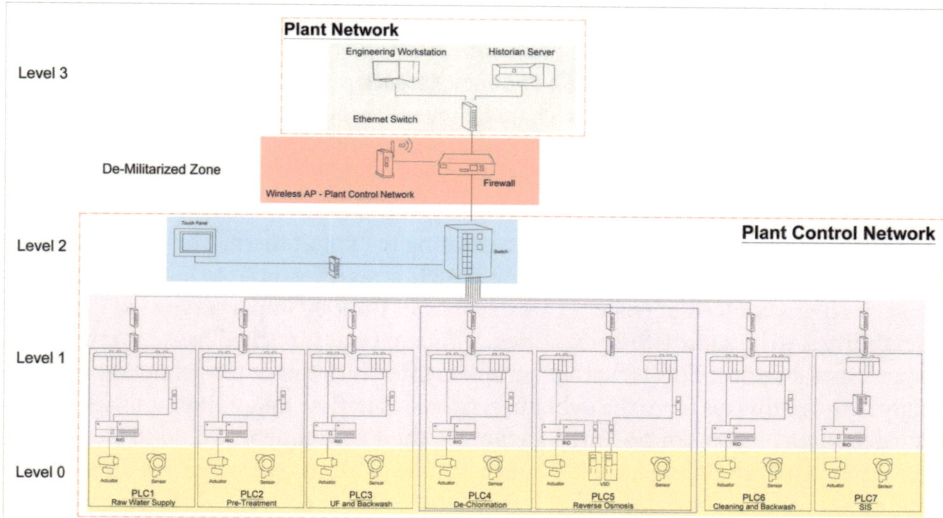

Figure 4.
SWaT network diagram zones and conduits.

5. Cyber risk quantification methodology

The second contribution of this chapter conceptualized how cyber risk can be quantified using existing methodologies advocated in the IT domain such as FAIR that is an open, international standard risk model designed to quantify risk backed by industry polling or cyber-incident disclosures to improve objectivity [3], and the concept of the Swiss cheese model that is a risk management framework from an analogy of slices of Swiss cheese stacked together to visualize how multiple layers of defense can prevent accidents. Through the lenses of these concepts, layers of defense, or barriers, can prevent the manifestation of impacts that transcend from the cyber domain to the physical world.

Risk analysis is typically a component within the larger framework of risk assessment. The purpose of analysis is to evaluate significance and compare options, providing accurate information for decision-making [19]. For an analysis to be valid, it requires objectivity, and the goal is to drive objectivity into the process and data, but it is important in the FAIR approach that all analysis has some degree of subjectivity. Any human interpretation of risk is never as absolute as the scientific laws.

Fundamentally, risk comprises loss magnitude (LM) and loss event frequency (LEF). LEF is the probable frequency at which loss will materialize from a threat agent's action within a given timeframe. LM is the probable magnitude of loss resulting from an event. In this chapter, LM refers to the significant loss of containment selected from an ALARP approach in the case study. The PFD or likelihood of the occurrence caused by a cyber-incident that results in a relevant safety incident must be manifested from a failure of the safeguards upon demand. Therefore, LEF is the PFD of either category of elimination and engineering controls in **Table 2**, that is, AIT402 (ORP) High and $NaHSO_3$ dosing system, and AIT504 (Conductivity) High, and MV501 Open and MV503 Close, to perform its role as an independent safeguard under a specified risk scenario.

Loss event frequency (LEF) is divided into threat event frequency and vulnerability. Threat event frequency is simply the likely rate at which threat agents will cause a loss within a set period. It is a combination of contact frequency (how often agents meet assets) and probability of action (the chance they will act on an asset after contact). Meanwhile, vulnerability is the probability that an agent's actions will result in an actual loss, and this is determined by their threat capability versus the difficulty they encounter.

Vulnerability is influenced by both the strength and effectiveness of existing controls and the capabilities of the threat agents attempting to exploit them. The type of threat actors, ranging from less sophisticated actors such as script kiddies to highly resourced adversaries like advanced persistent threats (APTs), directly impacts how likely a vulnerability is to be successfully exploited. This can be approached based on the threat actor and threat vector. This is further complemented by a threat modeling approach to analyze the pathways from the cyber in terms of data flow and vulnerabilities to the physical domain in terms of potential hazards to their consequences, and to identify the necessary control measures to prevent and mitigate these risks. The probability of a vulnerability resulting in loss is therefore dependent on the effectiveness of risk controls. In parallel, threat event frequency depends not only on the presence of a vulnerability but also on its exploitability in real-world environments. Tools such as the exploit prediction scoring system (EPSS) provide a dynamic measure of how likely a given vulnerability is to be exploited in the wild, thereby linking vulnerability management to the real-world threat landscape and enabling more informed risk prioritization [20].

Figure 5 shows the decomposition of risk into its risk factors and the corresponding objective risk quantification approaches. To illustrate how this risk quantification approach works, a relevant cybersecurity risk scenario is first proposed. The vulnerability is semi-quantified based on the effectiveness of controls in relation to the attack, whereas exploit prediction scoring system (EPSS) is a data-driven tool that estimates the likelihood a software vulnerability will be exploited. It does this by using current threat information from CVE and real-world exploit data. The LEF, that is the likelihood of loss materialization from the threat agent's action on the vulnerability can then be calculated from both threat event frequency and vulnerability.

Based on a report of cyber-attacks on European and US enterprises in 2022 [21], remote access server accounts for one of the top 5 initial entry points of cyber-attacks. To quantify the cyber risk for the proposed cybersecurity risk scenario, the

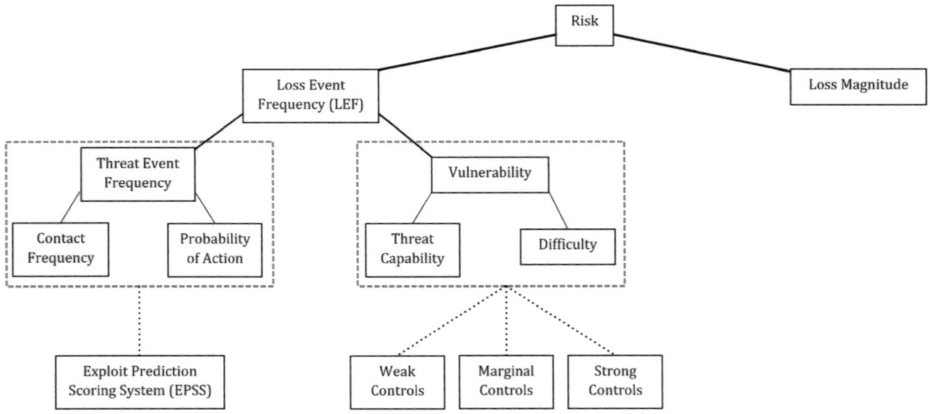

Figure 5.
Decomposition of risk factors.

semi-quantitative methodology assumes the perpetrator is an APT that uses malware to exploit remote access capability in the historian server to obtain administrator credentials and perform lateral movement to the Engineering Workstations and hence, PLCs 4 and 5 *via* the firewall. The historian server, being a database, is affected by the Log4Shell (CVE-2021-44228) that is designated by Mitre with the highest severity rating of 10.0 [22]. Log4j is a widely used open-source Java logging library found in many software applications and services. If exploited, this vulnerability could allow attackers to gain complete control of affected systems. Historian server is usually considered a crown jewel of asset identification, and administrator credentials in this server have a wide traversal capability in the north-south and east-west traffic. This can result in the same risk scenario whereby excess oxidant damages the final membrane barrier to cause a potential loss of containment risk scenario.

5.1 Vulnerability

The effectiveness of ISA 62443 standards can be assessed for key controls translated from a SL maturity scale. A possible subjective scale for SLs 0 and 1 is considered weak, SL 2 is deemed marginal, and SLs 3 and 4 are categorized as strong. The vulnerability contribution can be calculated from the control effectiveness, and this calculation increases in accuracy when a control assurance exercise is conducted using audit-grade sampling [3]. Weak, marginal, and strong controls are allocated 100%, 50% and 10% vulnerability ratings, respectively.

To evaluate an individual foundational requirement's role in mitigating overall scenario risk, control effectiveness can be weighted according to the control's specific type. The analytic hierarchy process offers a method for establishing these weightings. For the threat scenario described above, based on the APT threat actor,

a. FR-1. Identification and authentication control, FR-2. Use controls are key controls to stop the exploitation of vulnerability.

b. FR-5. Restricted data flow can prevent the malware from transversing from Level 3 to Level 1 of **Figure 5**.

c. FR-6. Timely response to events and FR-7. Resource availability are important controls to address the impact of the attack.

Therefore, weightage is the highest for FRs 1/2, followed by FR 5 and then FRs 6/7, with FRs 3/4 with the lowest weightage, that is, 0% since the controls are not relevant to the threat scenario. **Table 4** illustrates how a semi-quantitative method can approximate the overall vulnerability or likelihood of risk given existing controls, resulting in a quantified probability of 0.14 that a threat agent's actions will lead to a loss. As the vulnerability of the safeguards from either category of elimination and engineering controls in **Table 2**, that, AIT402 (ORP) High and $NaHSO_3$ dosing system, and AIT504 (Conductivity) High, and MV501 Open and MV503 Close, is similar based on the threat vector and data flow, the probability resulting in a loss for either of these controls under this risk scenario is the same.

The protection of critical control of either category of elimination and engineering controls in **Table 2**, that is, AIT402 (ORP) High and $NaHSO_3$ dosing system, and AIT504 (Conductivity) High, and MV501 Open and MV503 Close, at SL 4 is already assumed the conservative approach based on the operator failing to manually

Scenario	Key controls (IES 62443 foundation requirements)	Weight of controls (%)	Current control rating	Vulnerability (%)	Vulnerability contribution (%)	Overall vulnerability (%)
Malware exploiting remote access capability *via* the historian server using administrative credentials to cause plant disruption	FR-1. Identification and authentication control (IAC)	30	Strong	10	3	14
	FR-2. Use control (UC)	30	Strong	10	3	
	FR-3. System integrity (SI)	0	Weak	100	0	
	FR-4. Data confidentiality (DC)	0	Weak	100	0	
	FR-5. Restricted data flow (RDF)	20	Strong	10	2	
	FR-6. Timely response to events (TRE)	10	Marginal	50	5	
	FR-7. Resource availability (RA)	10	Strong	10	1	

Table 4.
Semi-quantified risk profile.

intervene based on the non-hackable alarm signals as a mitigation measure caused by a failure of automation due to cyber-interference. Hence, the lowest overall vulnerability must be limited to 10% due to the limitations of human error.

5.2 Threat event frequency

The EPSS dynamically estimates the chance a vulnerability will be exploited in the coming month, all based on data. Unlike static severity scores, EPSS continuously analyzes millions of data points, considering exploit availability, scanning activity, and threat intelligence to adjust its predictions over time. Each variable's influence can change depending on context. This predictive capability is particularly valuable in operational technology (OT) environments, where systems due to constant uptime operational requirements or end-of-life systems, are often difficult or impossible to patch.

During the Log4j incident, historian servers using vulnerable Java libraries became a hidden weak point. The early number of references to a vulnerability like Log4Shell (CVE-2021-44228) initially had a score of 0.355 meaning a 35.5% probability of exploitation in the wild, but as the publication and sharing of exploit code greatly increases the probability of observing exploitation, there was a strong relationship between exploit code being published in GitHub, Exploit DB and Metasploit. Observed exploitation activity resulted in an eventual updated score of 0.944. This indicated a 94.4% probability of exploitation of the Log4Shell vulnerability within the next 30 days [22].

EPSS does not aim for perfect foresight, but it consistently outpaces human analysts in sifting through massive datasets and uncovering patterns, offering a useful reference point for emerging threats. EPSS can help asset owners prioritize emergency mitigations by showing a high and rising probability of exploitation, even before wide-scale attacks are observed internally. Over time, EPSS scoring also reveals when the threat intensity begins to decline, helping organizations decide when urgent mitigations could be safely scaled back. In this way, EPSS offers a pragmatic tool to align defensive actions with real-world threat behavior, rather than theoretical risk alone.

5.3 Loss event frequency (LEF)

Based on the proposed cybersecurity risk scenario of a malware exploiting remote access capability *via* the historian server using administrative credentials to cause plant disruption by APTs, both PLCs 4 and 5 have to be compromised simultaneously to cause a potential loss of containment risk scenario for the LOPA described earlier in this chapter. Hence, the overall vulnerability of the system can be semi-quantified as $1.96 \times 10e^{-2}$. As EPSS for Log4Shell exploitation on the historian server is a range of 0.355 to 0.944, the overall LEF can be reasonably quantified to within a range of $1.85 \times 10e^{-2}$ to $6.95 \times 10e^{-3}$.

5.4 Cyber risk quantified PHA risk assessment

Table 5 shows the LOPA risk calculation for a cyber risk quantified approach. The mitigated event frequency resulting from a cyber-attack, calculated using the FAIR methodology as described in this chapter, will be $7.4 \times 10e^{-4}$ to $2.78 \times 10e^{-5}$. The

Safeguards		**Independent protection layers (IPLs)**	**Mitigated event frequency (no cyber-attack)**	**Cyber risk quantified mitigated event frequency (cyber-attack)**
Administrative control	AIT203 (ORP) High	Alarm 1—AIT203	0.2	0.2
	AIT302 (ORP) High	Alarm 2—AIT302	0.2	0.2
Elimination control	UV401 On	Dechlorinator 1 (UV401)	0.1	1
	AIT402 (ORP) High and $NaHSO_3$ dosing system	Dechlorinator 2 (*AIT402 and $NaHSO_3$ dosing system*)	0.1×0.1	$1.85 \times 10e^{-2}$–$6.95 \times 10e^{-3}$
	AIT502 (ORP) High	Interlock 3—AIT502 (RO shutdown)	0.1	1
Engineering control	FIT502 High	Alarm 3—FIT502	0.5	1
	AIT504 (Conductivity) High	Out-of-spec water (*input bypass*)	0.1×0.1	$1.85 \times 10e^{-2}$–$6.95 \times 10e^{-3}$
	MV501 Open and MV503 Close			
Total frequency[1]			$4 \times 10e^{-6}$	$7.4 \times 10e^{-4}$–$2.78 \times 10e^{-5}$

[1]IPLs have to be independent. Devices linked to the same PLC will not be double counted and use the lower of the mitigated risk frequency.

Table 5.
LOPA risk calculation using a cyber risk quantification methodology.

tolerable risk frequency of $1 \times 10e^{-4}$ determined for this case study can be met partially based on the cyber risk quantified approach that fits the concept of an ALARP situation. This is in comparison of the earlier recommendation to allocate the relevant safeguards with SL 4 without cyber risk quantification.

6. Inherent difference between risk and reliability

It is crucial to acknowledge the distinct nature of safety instrumented systems (SIS) within the broader framework. The risk-based approach in this chapter encompasses process safety to determine if there are sufficient safeguards/IPLs to reduce the risk of a particular scenario from occurring in the presence of a deliberate cyber-attack. On the contrary, SIS ensures the reliability of the current safeguards/IPLs to reduce the risk of the scenario further by ensuring the safety integrity levels through multiple, redundant components and more rigorous and management requirements [23]. The risk-based approach is not able to (semi-) quantify the reliability of the SIS when a cyber-attack happens.

IEC 61511 underscores the need for a separation of SIS from the plant network to safeguard against potential cyber threats [24]. In the event the plant network is compromised, this separation ensures that the integrity and functionality of the safety instrumented systems remain uncompromised, as they operate independently of the plant network. As part of the security risk assessment, risk-based security levels are then used to implement the SIS that is tailored based on the risk level determined by the management team of the plant. The integration of these security levels, in adherence to the IEC 62443 standards, provides a robust defense mechanism against cyberattacks, ensuring the resilience and reliability of safety instrumented systems.

By acknowledging and addressing this fundamental distinction between the risk-based approach applied to general safety considerations and the reliability-centric approach inherent in safety instrumented systems, this chapter will help readers grasp the intricate nature of safety and security within industrial settings. This distinction not only aligns with established standards and guidelines but also contributes to a more holistic and specialized discourse surrounding the protection of critical systems in the face of evolving cyber threats.

7. Conclusion

This chapter outlines a structured approach for operationalizing a safety risk quantification framework for cyber risk events, grounded in established industry standards. By leveraging existing practices from the safety lifecycle and adapting them through a risk-based lens, practitioners are provided with a familiar and objective method to address cybersecurity risks within the context of functional safety. The cyber risk quantification method illustrated is founded on the basis of objectivity and increases in value as statistical measurements become available, especially in the OT domain. The more economical and effective approach is to prioritize the cybersecurity controls in the relevant zones and conduits using a widely accepted LOPA safety methodology until statistical data for cyberattacks in the process industries becomes widely available. This bears resemblance to how various techniques to ensure comprehensive risk assessment and management in industrial and process settings such as HAZOP, LOPA, Swiss cheese model, can be delivered complementary with the traditional cybersecurity management system (CSMS) proposed by the IEC 62443 standards. The deliberate combination of these industry consensus-based approaches under a workshop setting, albeit currently carried out separately in different domains of security and safety, can help to ease the integration into existing practices that reduce the complexity and inertia for adoption.

Acknowledgements

This study was funded by Conny Tech Pte Ltd., Singapore and supported by the Enterprise Singapore Industrial Post-Graduate Programme Grant (25-IPPIII-AM-001-01).

Breviary

ALARP—as low as reasonably practicable:	Risk management principle requiring risks to be reduced to the lowest level reasonably achievable
AIT—analyzer indicating transmitter:	Instruments that measure, indicate, and transmit process variables
APT—advanced persistent threat:	Sophisticated, targeted cyber-attack carried out by skilled adversaries
Conduits:	Communication pathways or sets of communication assets with common security requirements
Contact frequency:	How often threat agents encounter target assets
CSMS—cybersecurity management system:	Systematic approach to managing cybersecurity risks in industrial control systems
CVE—common vulnerabilities and exposures:	Database of publicly disclosed cybersecurity vulnerabilities
EPSS—exploit prediction scoring system:	Data-driven tool that estimates the likelihood a software vulnerability will be exploited
FAIR—factor analysis of information risk:	Quantitative risk analysis framework for cyber risk assessment
FIT—flow indicating transmitter:	Instruments that measure and transmit flow rates
HAZOP—hazard and operability study:	Structured methodology to identify process hazards and operational issues through systematic examination of process deviations
HMI—human machine interface:	User interface that connects operators to machines/systems
ICS—industrial control systems:	Computer-controlled systems used to monitor and control industrial processes

IE—initiating events:	Events that can lead to an undesired consequence if protection layers fail
IEC 61511:	International standard for functional safety of safety instrumented systems in process industries
IPL—independent protection layer:	A device, system, or action that reduces risk independently of the initiating event and other protection layers
ISA-84.00.01-2004:	Safety instrumented systems standard for the process industries
ISA-TR84.00.09-2023:	Technical report on cybersecurity related to functional safety lifecycle
LEF—loss event frequency:	The probable frequency that loss will materialize from a threat agent's action within a given timeframe
LM—loss magnitude:	The probable magnitude of loss resulting from an event
LOPA—layer of protection analysis:	Semi-quantitative risk assessment method that determines if sufficient independent protection layers exist to achieve tolerable risk
Membrane barrier:	Physical barrier in water treatment systems that can be damaged by excess oxidants
MV—motor valve:	Motorized valves controlled by the automation system
$NaHSO_3$—sodium bisulfite:	Chemical used for dechlorination in water treatment
ORP—oxidation-reduction potential:	Measurement of water's ability to oxidize or reduce contaminants
OT—operational technology:	Hardware and software that monitors and controls physical devices, processes, and infrastructure
P&ID—piping and instrumentation diagram:	Detailed schematic illustration that shows the piping, equipment, instrumentation, and control interlocks for a process system. It serves as a key document in engineering design, process safety, operations, and maintenance
PCAI—process control, alarms, and interlocks:	Core control system components that monitor and control industrial processes

DOI: http://dx.doi.org/10.5772/intechopen.1011612

PFD—probability of failure on demand:	The probability that a safety system will fail to perform its intended function when required
RE—requirement enhancements:	Additional measures that may be needed to meet higher security levels
SL-A—security level achieved:	Actual security level attained by a system after implementation
SL-C—security level capability:	Security level that system components can inherently support when properly configured
SL-T—security level target:	Intended security level for a specific system, determined through risk assessment
SIS—safety instrumented system:	Independent system designed to bring a process to a safe state when dangerous conditions occur
SUC—system under consideration:	The specific system, zone, or conduit being analyzed for security requirements
SR—system requirements:	Specific requirements that must be met to achieve foundation requirements
PHA—process hazard analysis:	Systematic study to identify potential hazards and operability problems in industrial processes
PLC—programmable logic controller:	Industrial computer used to control manufacturing processes
RO—reverse osmosis:	Water purification technology using semipermeable membranes
TEF—threat event frequency:	Rate at which threat agents cause loss events within a specified timeframe
Threat capability:	Skills, resources, and motivation of threat agents
UV—ultraviolet:	UV disinfection system component
Ultrafiltration:	Membrane filtration process that removes particles and pathogens
Vulnerability:	Probability that a threat agent's actions will result in loss
QRA—quantitative risk analysis:	Numerical assessment of risk using statistical methods
Swiss cheese model:	Risk management framework visualizing how multiple defense layers prevent accidents
Zones:	Logical or physical groupings of assets with shared security requirements

Author details

Victor Siang Tze Sim[1*], Kok Hwa Lim[2] and Tan Yong Rui[1]

1 Conny Tech Pte Ltd, Singapore

2 Singapore Institute of Technology, Singapore

*Address all correspondence to: victor.sim@conny.sg

References

[1] Sim STV, Chuadhry MA, Koh YKK, Zhou J. Safety watermark: A defense tool for real-time digital forensic incident response in industrial control systems. In: Applied Cryptography and Network Security Workshops, ACNS 2023, Lecture Notes in Computer Science. Germany: Springer; 2023. pp. 299-320. DOI: 10.1007/978-3-031-41181-6_17

[2] Boehm J, Curcio N, Merrath P, Shenton L, Stähle T. The Risk-Based Approach to Cybersecurity. United States: McKinsey & Company; 2019. Available from: https://www.mckinsey.com/capabilities/risk-and-resilience/our-insights/the-risk-based-approach-to-cybersecurity [Accessed: April 29, 2025]

[3] Vohradsky D. The cyberrisk quantification journey. ISACA Journal. 2022;**2**:19-32. Available from: https://www.isaca.org/resources/isaca-journal/issues/2022/volume-2/the-cyberrisk-quantification-journey [Accessed: April 29, 2025]

[4] ISA-TR84.00.09-2024. Cybersecurity Related to the Functional Safety Lifecycle. United States: International Society of Automation; 2024

[5] Morris M. How Cyber-Informed Engineering Can Be the Way Forward for Critical Infrastructure [Internet]. United States; 2022. Available from: https://www.forbes.com/sites/forbestechcouncil/2022/11/09/how-cyber-informed-engineering-can-be-the-way-forward-for-critical-infrastructure/?sh=1a981a966614 [Accessed: April 29, 2025]

[6] ANSI/ISA 61511-2017. Functional Safety: Safety Instrumented Systems for the Process Industry Sector—Part 1: Framework, Definitions, System, Hardware and Software Requirements. United States: International Society of Automation; 2017

[7] ANSI/ISA 62443-3-2-2020. Security for Industrial Automation and Control Systems, Part 3-2: Security Risk Assessment for System Design. United States: International Society of Automation; 2020

[8] ANSI/ISA 62443-4-2-2018. Security for Industrial Automation and Control Systems, Part 4-2: Technical Security Requirements for IACS Components. United States: International Society of Automation; 2018

[9] ANSI/ISA 62443-3-3-2013. Security for Industrial Automation and Control Systems, Part 3-3: System Security Requirements and Security Levels. United States: International Society of Automation; 2013

[10] ANSI/ISA 62443-1-1-2007. Security for Industrial Automation and Control Systems, Part 1-1: Terminology, Concepts, and Models. United States: International Society of Automation; 2007

[11] Tantawy A, Abdelwahed S, Erradi A. Cyber LOPA: An integrated approach for the design of dependable and secure cyber-physical systems. IEEE Transactions on Reliability. 2002;**71**:1075-1091. DOI: 10.1109/TR.2022.3163652

[12] Casey R. Limitations and misuse of LOPA. Loss Prevention Bulletin. 2019;**265**:13-16. Available from: https://www.icheme.org/media/8735/lpb265_pg13.pdf [Accessed: April 29, 2025]

[13] Sim STV. Safety watermark—Defense tool for real time digital forensic incident response [thesis]. Singapore: Information Systems Technology and Design:

Singapore University of Technology and Design; 2023

[14] Adepu S, Mathur A. Distributed attack detection in a water treatment plant: Method & case study. IEEE Transactions on Dependable and Secure Computing. 2021;**18**:86-99. DOI: 10.1109/TDSC.2018.2875008

[15] Chuadhry MA, Venkata RP, Vishrut KM. A practical physical watermarking approach to detect replay attacks in a CPS. Journal of Process Control. 2022;**116**:136-146. DOI: 10.1016/j.jprocont.2022.06.002

[16] Environmental Public Health Act (Chapter 95) (2019), Environmental Public Health (Water Suitable for Drinking) (No. 2) Regulations 2019 in Singapore Statutes Online [Internet]. 2019. Available from: https://sso.agc.gov.sg//SL/EPHA1987-S274-2019?DocDate=20190401 [Accessed: April 29, 2025]

[17] Workplace Safety and Health—Major Hazard Installation. ALARP Demonstration Guidelines: Single Scenario Risk Tolerability Target and Adequacy of Barriers in Ministry of Manpower, Singapore [Internet]. 2020. Available from: https://www.mom.gov.sg/workplace-safety-and-health/major-hazard-installations/preparing-for-safety-case [Accessed: April 29, 2025]

[18] Corso PS, Kramer MH, Blair KA, Addiss DG, Davis JP, Haddix AC. Cost of illness in the 1993 waterborne cryptosporidium outbreak, Milwaukee, Wisconsin. Emerging Infectious Disease. 2003;**9**:426-431. DOI: 10.3201/eid0904.020417

[19] Freund J, Jones J. Measuring and Managing Information Risk: A FAIR Approach. United States: Butterworth-Heinemann; 2014. ISBN-13: 978-0124202313

[20] Jacobs J, Romanosky S, Edwards B, Roytman M, Adjerid I. Exploit Prediction Scoring System (EPSS). United States: Blackhat; 2019. Available from: https://i.blackhat.com/USA-19/Thursday/us-19-Roytman-Predictive-Vulnerability-Scoring-System-wp.pdf [Accessed: April 29, 2025]

[21] Wharton G. Cyber Readiness Report 2022. Bermuda: Hiscox; 2022. Available from: https://oneclick-cloud.com/wp-content/uploads/2023/09/Hiscox-Cyber-Readiness-Report-2022.pdf [Accessed: April 29, 2025]

[22] Jacobs J. EPSS vs Log4Shell. Forum of Incident Response and Security Teams, Inc Blackhat [Internet]. 2022. Available from: https://www.first.org/epss/articles/log4shell [Accessed: April 29, 2025]

[23] Center for Chemical Process Safety—CCPS. What is a Safety Instrumented System? United States: American Institute of Chemical Engineers—AIChE; 2009. Available from: https://www.aiche.org/sites/default/files/2009-07-Beacon-English.pdf [Accessed: April 29, 2025]

[24] Timms C, Kirkwood D. The Principles of IEC 61508 and IEC 61511—Functional Safety Instrumented Systems for the Process Industry Sector (Ed 2.0 of IEC 61511), TüV Rheinland Functional Safety Engineer Program. United Kingdom: C&C Technical Support Services; 2022

Chapter 5

Calculation of Stability Domains for Process Control Systems with PI Controllers, Based on the Hurwitz Criterion

Constantin Volosencu

Abstract

The chapter falls within the broad context of the study of the stability of continuous-time linear control systems and the design of related PI controllers. The study presents a synthesis of the stability domains of PI controller parameters in the case of second- and third-order process control systems and analyzes the specific choice of these parameters. In determining the stability domains, the Hurwitz stability criterion is used. The paper has a didactic, illustrative character, presenting the domains of stability analytically and graphically, with case studies, simulations and examples. Several values are taken for the controller parameters, from the determined domains, and the frequency characteristics of the open-loop system and the unitary step signal responses of the closed loop are determined, which test the calculated stability domain.

Keywords: control systems, PI controllers, internal stability, Hurwitz criterion, system modeling and simulation

1. Introduction

Control theory is an applied mathematics on control of dynamical systems, represented in practice by processes and machines, electric drives, the flow of water in a pipe, temperature in a room and so on [1]. Control theory is working with mathematical models as ordinary differential equations, transfer function and state space models. The researchers in the field of control theory are developing algorithms to control these systems to bring the process variables to the desired value as a set point. In this theory, some aspects which are studied are stability, controllability and observability, concepts which characterized the control systems. The researchers are designing automation that improve manufacturing and all kinds of equipment [2–4]. The stability theory has as its object the study of the stability of solutions to differential equations and dynamical systems. For continuous time systems the model of autonomous nonlinear dynamical systems is taken in consideration. In the control system theory, the Hurwitz stability criterion is a mathematical test that is a necessary and sufficient condition for the stability of a linear time-invariant dynamical system. Hurwitz criterion arranges the coefficients of the polynomial into a square matrix, called the Hurwitz matrix, and showed that the polynomial is stable if and only if the sequence of determinants of its principal submatrices are all positive. The second-order and the third-order models are frequently used in theory to describe processes in the design of automatic control systems. The PI controller is also a type often used in practice [5, 6]. The design phases of a linear control system are analysis, modeling, stability analysis, synthesis, controller tuning, frequency analysis and time domain analysis. Methods to design proportional-integral controllers are presented in literature, as in [7], based on the determination of the set of stabilizing PI controller with closed loop poles having real parts negative, to obtain the set in the controller parameter space. The chapter [8] discusses the stability of closed-loop systems, based on the Routh-Hurwitz stability criterion and the root locus, with influence on the closed-loop plant. In the article [9], a study on the topic is presented, in which the relationship depending on the parameters of the PI regulator is linear, and the Routh criterion is used to determine the stability domain. The article [10] discusses the application of the Routh-Hurwitz criterion for stability analysis on the design of PI controllers, with examples. The Hurwitz test is classical. Graphical representation of stability regions in (K_R, T_R) space of the controller parameters for the second order process has been used before in literature, in the above works.

The present chapter presents a study of continuous-time linear control systems and the design of related PI controllers. The study presents a synthesis of the stability domains of PI controller parameters in the case of second and third-order process control systems and analyzes the specific choice of these parameters, based on a standard Hurwitz-based approach. The chapter has a didactic, illustrative character, presenting the domains of stability analytically and graphically, and not as a novel methodology, and with case studies, simulations and examples. Several values are taken for the controller parameters, from the determined domains, and the frequency characteristics of the open-loop systems and the unitary step signal responses of the closed loop systems are determined, which test the calculated stability domain. The presented calculations constitute a theoretical basis, with a teaching role, in the design of linear control systems. The study can be extended to processes of order 4, 5 and so on.

2. Second order process

2.1 Control system structure

Consider a control system with the block diagram in **Figure 1**.

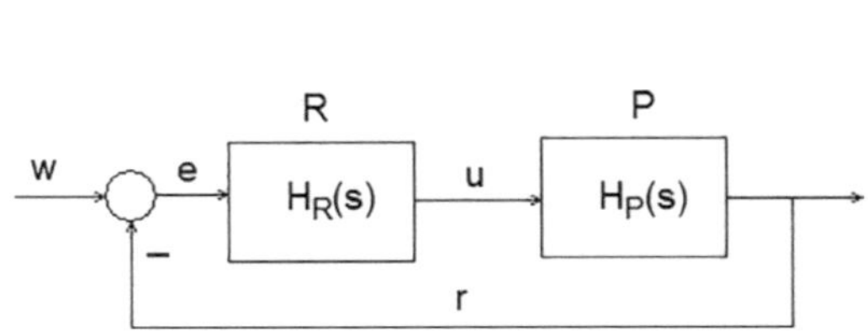

Figure 1.
Block diagram of the control system.

where R is the controller, P is the process, w is the control input signal, e is the error, u is the command input variable and r is the feedback. The transfer functions of the controller R and the process P of second order are $H_R(s)$ and $H_P(s)$, respectively:

$$H_R(s) = K_R\left(1+\frac{1}{sT_R}\right) \tag{1}$$

$$H_P(s) = \frac{K_P}{(a_2s^2+a_1s+1)} \tag{2}$$

Relation (1) contains the K_R and T_R parameters of the PI controller, which are the subject of design based on the stability domain that is deduced in the paper. Relation (2) contains the parameters a_1, a_2 and K_P of the second-order process, which determine the stability domain.

The transfer function of the closed-loop control system $w \rightarrow r$ from **Figure 1**, $r(s) = H(s)w(s)$ is:

$$H(s) = \frac{r(s)}{w(s)} = \frac{H_R(s)H_P(s)}{1+H_R(s)H_P(s)} \tag{3}$$

and it has the characteristic polynomial $P(s)$:

$$P(s) = a_2T_Rs^3 + a_1T_Rs^2 + T_R(1+K_RK_P)s + K_RK_P \tag{4}$$

The characteristic polynomial is of degree three, so the Hurwitz determinant is also of degree three.

In theory and practice, the process and controller coefficients a_1, a_2, K_P, K_R and T_R are positive.

Other relations, in various forms, can be used for mathematical models of the process and the controller. The calculations can be redone for other expressions for these models.

2.2 Application of the Hurwitz criterion

The Hurwitz determinant of the characteristic polynomial is

$$H = \begin{vmatrix} a_1T_R & K_RK_P & 0 \\ a_2T_R & T_R(1+K_RK_P) & 0 \\ 0 & a_1T_R & K_RK_P \end{vmatrix} \tag{5}$$

It is calculated by placing the corresponding coefficients of the characteristic polynomial $P(s)$ on the rows and columns.

The Hurwitz test is classical [7, 9, 10]. According to the Hurwitz criterion, the closed-loop control system with the characteristic polynomial $P(s)$ is stable if and only if all the coefficients of the polynomial are positive, what translates into relationships (6):

$$2T_R > 0, a_1T_R > 0, T_R(1+K_RK_P) > 0, K_RK_P > 0 \tag{6}$$

and the Hurwitz determinant and all its minors are positive, what translate into relations (7)–(9):

$$H > 0 \tag{7}$$

$$a_1T_R > 0 \tag{8}$$

$$a_1T_R^2(1+K_RK_P) - a_2T_RK_RK_P > 0 \tag{9}$$

From the relations above it follows that all the coefficients of the control system should be positive and only relation (9) is independent and will be used further in the analysis.

2.3 Stability domain

Starting from relation (9), the stability domain may be determined in the parameter plane (K_R, T_R) and also (T_R, K_R).

2.3.1 Stability domain in the parameter plane (K_R, T_R)

Rearranging Eq. (9) yields inequality (10):

$$T_R > \frac{a_2 K_R K_P}{a_1(1 + K_R K_P)} \tag{10}$$

where a_1, a_2, K_R and K_P are positive, which allows the determination of the stability domain.

By plotting the function:

$$f_1(K_R) = \frac{a_2 K_R K_P}{a_1(1 + K_R K_P)} \tag{11}$$

as in **Figure 2**.

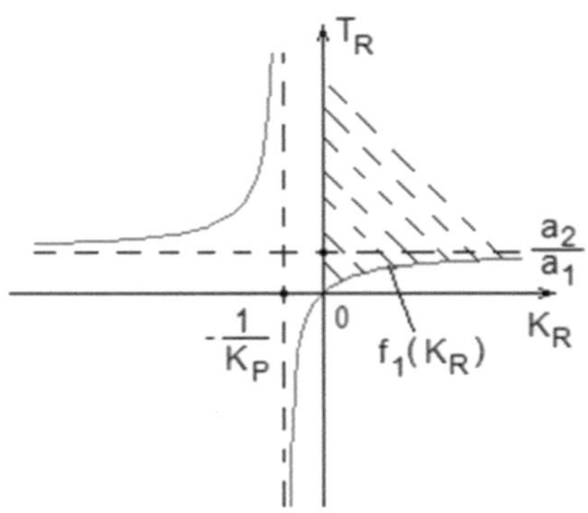

Figure 2.
Stability domain graph in the (K_R, T_R) plane.

the stability domain shaded in **Figure 2** is obtained.

Graphical representation of stability regions in (K_R, T_R) space has been used before in literature [7, 9, 10].

2.3.2 Stability domain in the parameter plane (T_R, K_R)

To determine the stability domain in coordinates (T_R, K_R), the relationship (9) is rearranged to obtain the function $f_2(T_R)$:

$$f_2(T_R) = \frac{a_1 T_R}{K_P(a_2 - a_1 T_R)} \tag{12}$$

by passing the coefficient K_R to the left and the coefficients with T_R to the right, in two situations.

The first situation is when the expression containing T_R is positive:

$$\begin{aligned} a_2 - a_1 T_R > 0 \\ T_R < \frac{a_2}{a_1} \end{aligned} \tag{13}$$

than

$$K_R > \frac{a_1 T_R}{K_P(a_2 - a_1 T_R)} \tag{14}$$

The second situation is when the expression containing T_R is negative:

$$\begin{aligned} a_2 - a_1 T_R < 0 \\ T_R > \frac{a_2}{a_1} \end{aligned} \tag{15}$$

than

$$K_R < \frac{a_1 T_R}{K_P(a_2 - a_1 T_R)} \tag{16}$$

By plotting the function $f_2(T_R)$ as in **Figure 3**:

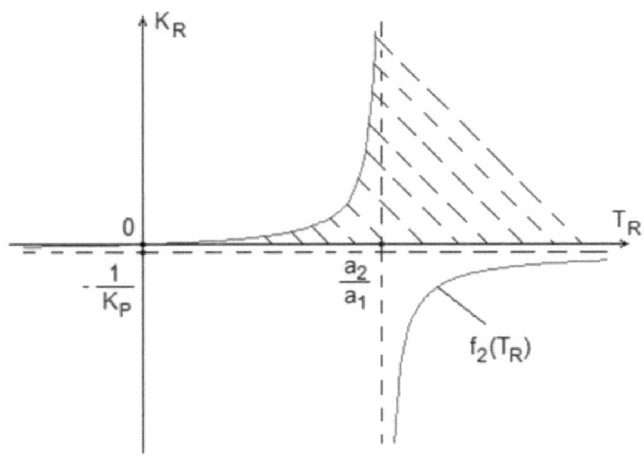

Figure 3.
Stability domain graph in the (T_R, K_R) *plane.*

the stability domain shaded in **Figure 3** is obtained, according to the relations (14) and (16).

2.4 Guidelines for choosing controller parameters

The graphs in **Figures 2** and **3** are two ways to illustrate the stability range for the PI controller parameters in the case of controlling a second-order process.

It can be seen that T_R can have values smaller positive than a_2/a_1 only for small positive values of K_R. And for large values of K_R, T_R must necessarily take values larger than a_2/a_1.

In practice, the time constant of the controller T_R is chosen to be correlated with the time constants of the process T_P. And then, using the phase reserve criterion, the open-loop amplification of the control system is chosen so that the control system ensures a sufficient phase margin, by choosing the controller amplification coefficient K_R. So, if $T_R = kT_P$, with the relation (16), K_R results in the range:

$$K_R < \frac{a_1 k T_P}{K_P(a_2 - a_1 k T_P)} \tag{17}$$

or in the range given by (14).

2.5 Case studies

Next, several case studies are presented to illustrate how the open-loop and closed-loop control system behaves for various values of the controller parameters chosen in various regions of the stability domain.

The following is an example calculation for the process with transfer function:

$$H_P(s) = \frac{1}{sT_1 + 1} \cdot \frac{1}{sT_2 + 1} \tag{18}$$

with the parameters: T_1 = 10 ms, T_2 = 1 ms, $a_1 = 10^{-6}$, $a_1 = 11.10^{-3}$, $K_P = 1$, $a_2/a_1 \approx 0{,}363.10^{-3}$, $-1/K_P = -1$.

In this case:

$$\begin{aligned} a_2 &= T_1 T_2, \\ a_1 &= T_1 + T_2 \end{aligned} \tag{19}$$

And the Eqs. (10) and (16) may be calculated according to the above relations.

We perform some frequency analyses of the open-loop control system and some time domain analyses in response to the input signal applied repeatedly, for some pairs of values chosen from the ranges in the two **Figures 2** and **3**: (a) T_R = 0.1 ms $< a_2/a_1$ and K_R = 5, (b) T_R = 1 ms and K_R = 5, (c) T_R = 0.76 ms $< a_2/a_1$ and K_R = 5.

The calculations were made using the MATLAB program [11]. The Appendix presents a MATLAB program that can be used to perform these analyses in the case of the second-order process. The numerical datasets used in the next case studies may be inserted in this program.

The amplitude and frequency characteristics are shown in **Figures 4–6**. The response characteristics to a unitary step signal are shown in **Figures 7–9**.

From the frequency characteristics it can be seen that the phase degree for zero magnitude is −226°. So, the phase margin is approximately −46°. The system is unstable.

It is observed that the system output signal obtained in response to the step input signal increases over time.

From the frequency characteristics, it can be seen that the phase margin is approximately 8°. The system is stable.

The system has a damped periodic response.

From the frequency characteristics it can be seen that the phase margin is approximately 1°. The system is stable.

The system has a periodic response with high oscillations.

A larger phase margin indicates a more stable system. A commonly accepted value is 45°. The ideal margin varies depending on the application, so on the values of the controller parameters. A phase margin of less than 45° may cause oscillation, while an excessively large margin can result in sluggish performance.

In a fourth study case the controller parameters are calculated in the optimal case with the Kessler modulus method, using the following relations:

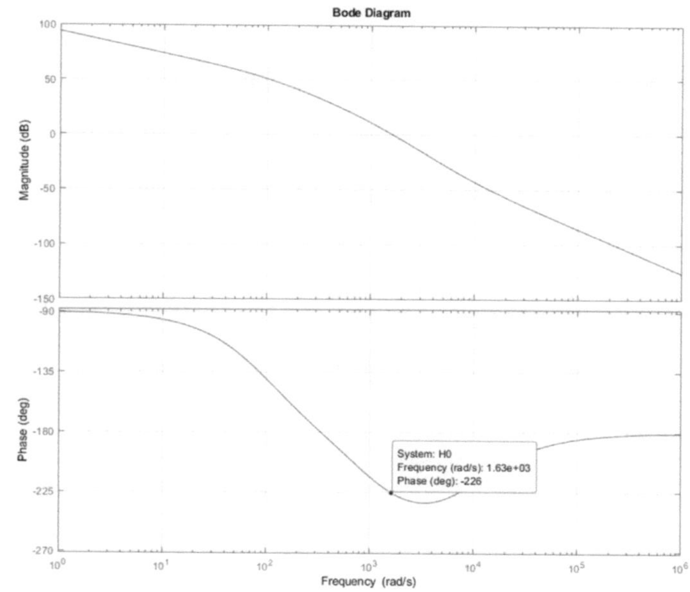

Figure 4.
Frequency characteristics for T_R = *0.1 ms and* K_R = *5.*

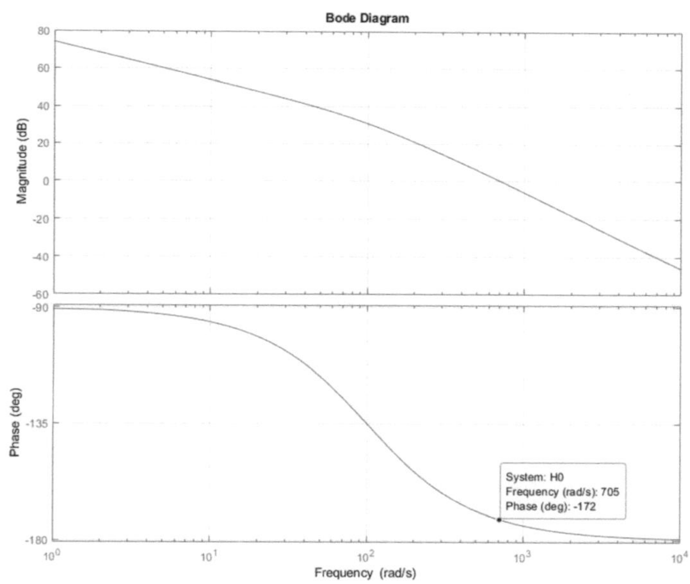

Figure 5.
Frequency characteristics for T_R = *1 ms and* K_R = *5.*

$$T_R = T_1 = 10\ ms, \quad K_R = \frac{T_1}{2K_P T_2} = 5 \tag{20}$$

The amplitude and frequency characteristics are shown in **Figure 10**. The response characteristic to a unitary step signal is shown in **Figure 11**.

From the frequency characteristics it can be seen that the phase margin is approximately 66° degrees. The system is stable.

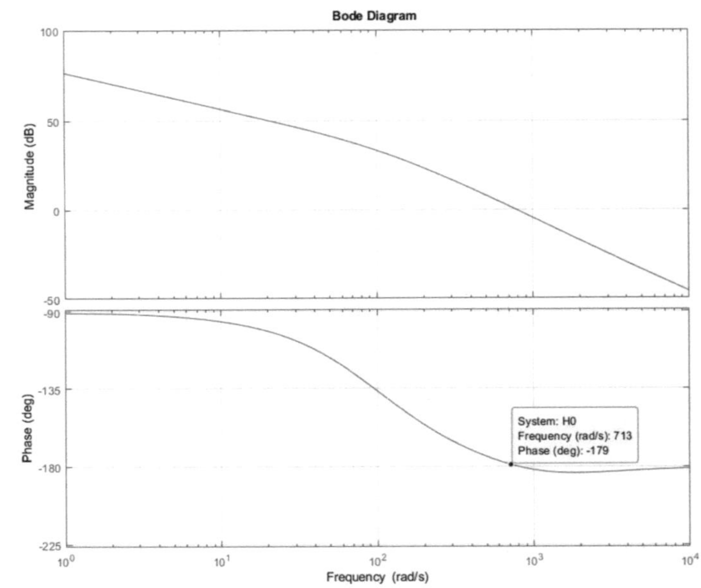

Figure 6.
Frequency characteristics for T_R = 0.76 *ms and* K_R = 5.

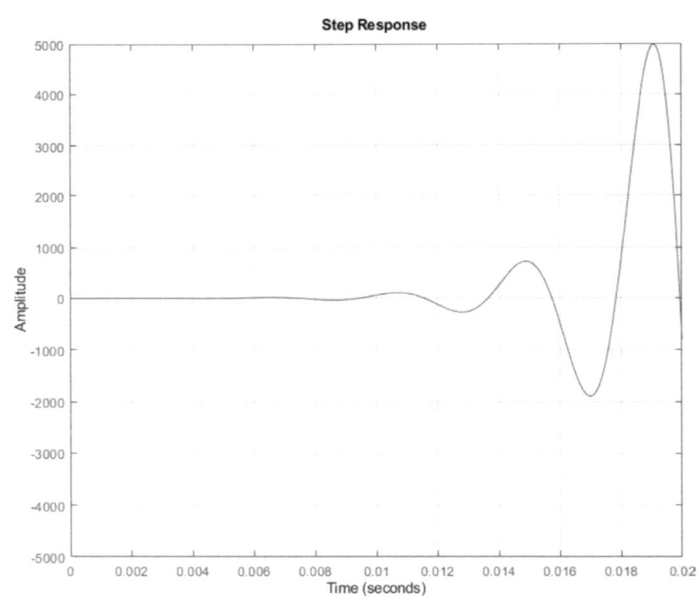

Figure 7.
Step response characteristic for T_R = 0.1 *ms and* K_R = 5.

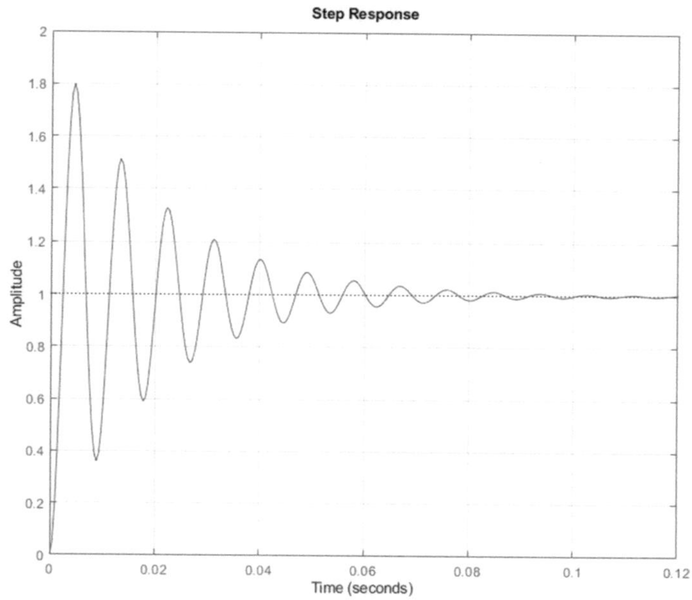

Figure 8.
Step response for $T_R = 1$ *ms and* $K_R = 5$.

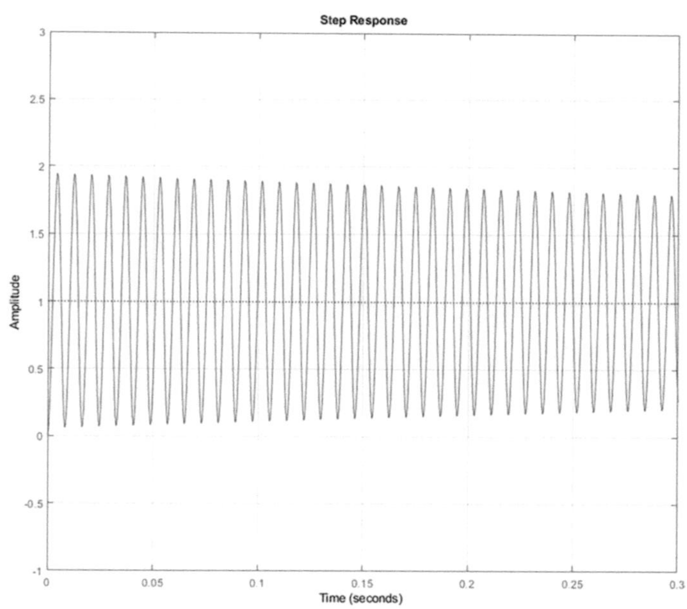

Figure 9.
Step response for $T_R = 0.76$ *ms and* $K_R = 5$.

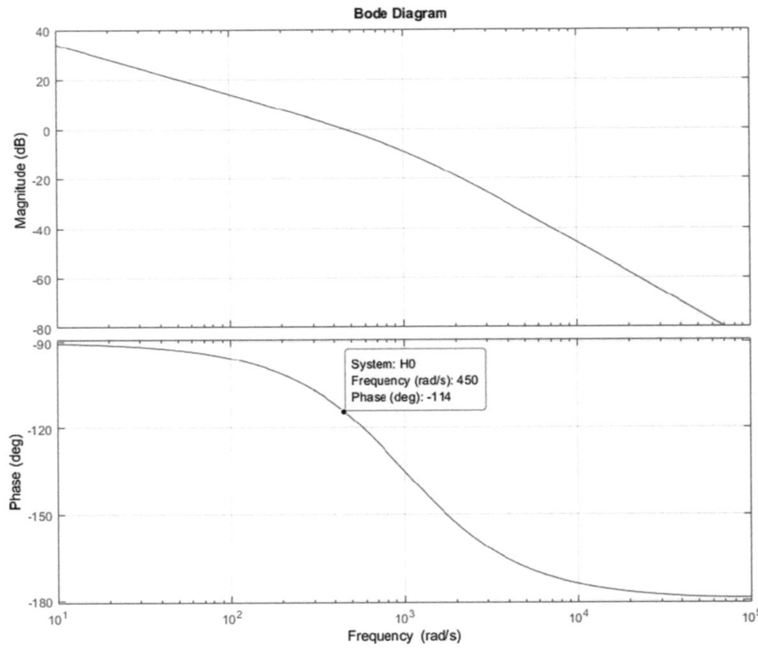

Figure 10.
Frequency characteristics for design with Kessler criterion.

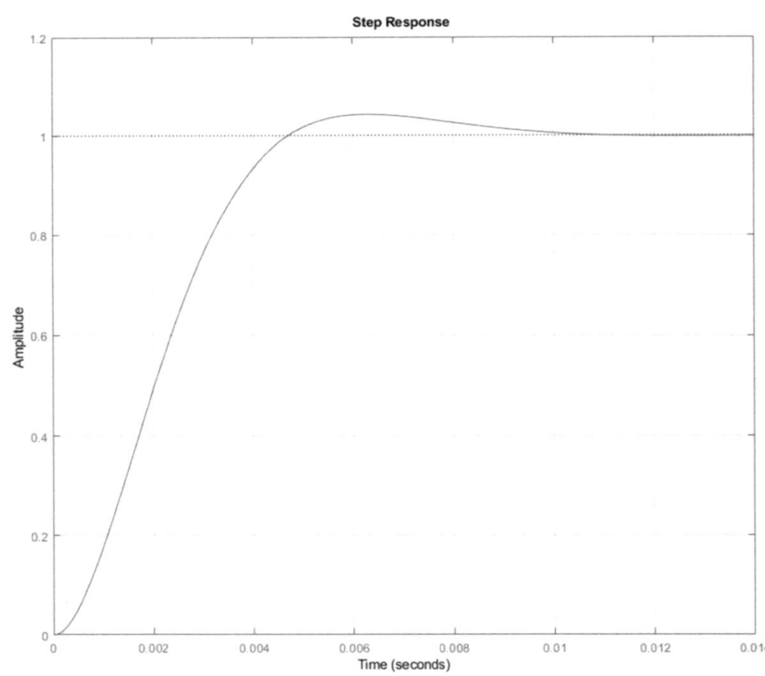

Figure 11.
Step response for design with Kessler criterion.

The values of the performance criteria that are obtained are: $\sigma_1\% = 4.3\%$; $t_1 = 4.7, T_2 = 4.7$ ms, $t_r = 8.4$ ms, $T_2 = 8.4$ ms.

3. Third order process

3.1 The control system

We consider the transfer function of the third order process given by the Eq. (20):

$$H_P(s) = \frac{K_P}{(a_3s^3 + a_2s^2 + a_1s + 1)} \tag{21}$$

with $K_P > 0, a_3 > 0, a_2 > 0, a1 > 0$, and the transfer function of the control system obtained with Eq. (3). The control system from **Figure 1** with the PI controller with the Eq. (1) has the transfer function calculated with Eq. (2) and the characteristic polynomial of the control system is

$$P(s) = a_3T_Rs^4 + a_2T_Rs^3 + a_1T_Rs^2 + T_R(1 + K_RK_P)s + K_RK_P \tag{22}$$

Since the characteristic polynomial has order four, the Hurwitz determinant will have also order four.

3.2 Stability domain

The Hurwitz determinant H of the characteristic polynomial is

$$H = \begin{vmatrix} a_2T_R & T_R(1 + K_RK_P) & 0 & 0 \\ a_3T_R & a_1T_R & K_RK_P & 0 \\ 0 & a_2T_R & T_R(1 + K_RK_P) & 0 \\ 0 & a_3 & a_1T_R & K_RK_P \end{vmatrix} \tag{23}$$

According to the Hurwitz criterion, the closed-loop control system with the characteristic polynomial $P(s)$ is stable if and only if all the coefficients of the polynomial are positive, what is transposed into relationships (24):

$$a_3 > 0, a_2T_R > 0, a_1T_R > 0, T_R(1 + K_RK_P) > 0, K_RK_P > 0 \tag{24}$$

and the Hurwitz determinant and all its minors are positive, what is transposed into relations (25) ÷ (28):

$$H_1 = a_2T_R > 0 \tag{25}$$

$$H_2 = a_1a_2T_R^2 - a_3T_R^2(1 + K_RK_P) > 0 \tag{26}$$

$$H_3 = -a_2^2T_R^2K_RK_P + T_R(1 + K_RK_P)H_2 > 0 \tag{27}$$

$$H = K_RK_PH_3 > 0 \tag{28}$$

From relations (23) and (24), it follows that all the coefficients of the controller and the process should be positive.

From relation (25), the relationship for determining the domain for K_R results:

$$K_R < K_{RM} = \frac{1}{K_P}\left(\frac{a_1 a_2}{a_3} - 1\right) \tag{29}$$

And

$$\frac{a_1 a_2}{a_3} > 1 \tag{30}$$

From relations (26) and (27), the relationship for determining the domain for T_R results:

$$T_R > \frac{a_2^2 K_P K_R}{(1 + K_P K_R)[a_1 a_2 - a_3(1 + K_P K_R)]} \tag{31}$$

The domain of stability results from the relations (28) and (30). The stability domain can be determined on the graph of the function $f(K_R)$:

$$f(K_R) = \frac{a_2^2 K_P K_R}{(1 + K_P K_R)[a_1 a_2 - a_3(1 + K_P K_R)]} \tag{32}$$

in the coordinate plane (K_R, T_R) from **Figure 12**.

3.3 Guidelines for choosing controller parameters

The relation (29) shows an obvious thing, namely, that the amplification coefficient K_R of the regulator must be smaller than a maximum value, so as to ensure the stability of the closed-loop control system.

And, the relation (31) shows an obvious thing, namely that the integration time constant TR of the regulator must be greater that a minimum value, so as to ensure the stability of the closed-loop control system.

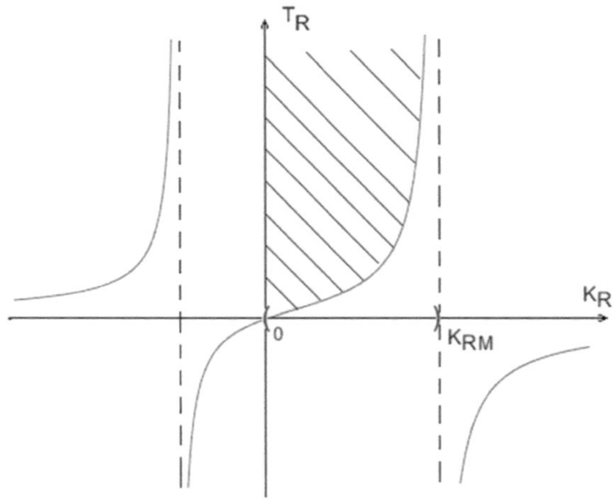

Figure 12.
Stability domain.

A particular case in practice may be

$$H_P(s) = \frac{K_P}{(T_1 s + 1)(T_2 s + 1)(T_3 s + 1)} \tag{33}$$

for which the equivalence is made

$$T_2' = T_2 + T_3 \tag{34}$$

And in this case, the discussion is moved with relation (18) to the case of the second-degree process.

The study can go from the case of the third-order process to the second order if $a_3 = 0$. Thus, the relation (28) becomes:

$$K_R < +\infty \tag{35}$$

And, the relation (30) becomes (10). In this case, the stability domain is the one obtained in the previous section.

3.4 Case studies

The following are example calculations for the process with the transfer function:

$$H_P(s) = \frac{1}{sT_1 + 1} \cdot \frac{1}{sT_2 + 1} \cdot \frac{1}{sT_3 + 1} \tag{36}$$

with the parameters: $T_1 = 10$ ms, $T_2 = 1$ ms, $T_3 = 0.1$ ms, $a_3 = 10^{-9}$, $a_2 = 0.111.10^{-4}$, $a_1 = 11.1.10^{-3}$, $K_P = 1$, $K_{RM} = a_1.a_2/a_3 - 1 = 122.21$.

In the first case, the controller parameters are calculated in the optimal case with the Kessler modulus method, using the following relations:

$$T_R = T_1 = 10\ ms, \quad K_R = \frac{T_1}{2K_P T_2'} \sim 4.545 \tag{37}$$

The calculation of the frequency characteristics of the open-loop system and the closed-loop system response can be done with the program in the Appendix, after modifying the related transfer functions accordingly as for a third-order process and obtaining an extended derivation of the program. The data sets from the next cased studies may be inserted in the extended derivation program.

The amplitude and frequency characteristics are shown in **Figure 13**.

The phase reserve is around 66°, the control system is stable, and the parameters of the controllers are in the stability domain.

The response characteristic to a unitary step signal is shown in **Figure 14**.

The overshoot of the response is around 4%, specific to Kessler modulus criterion.

In the second case, the controller parameters are chosen in the stability domain, but close to the maximum gain:

$$T_R = T_1 = 10\ ms, \quad K_R = 100 \tag{38}$$

The amplitude and frequency characteristics are shown in **Figure 15**.

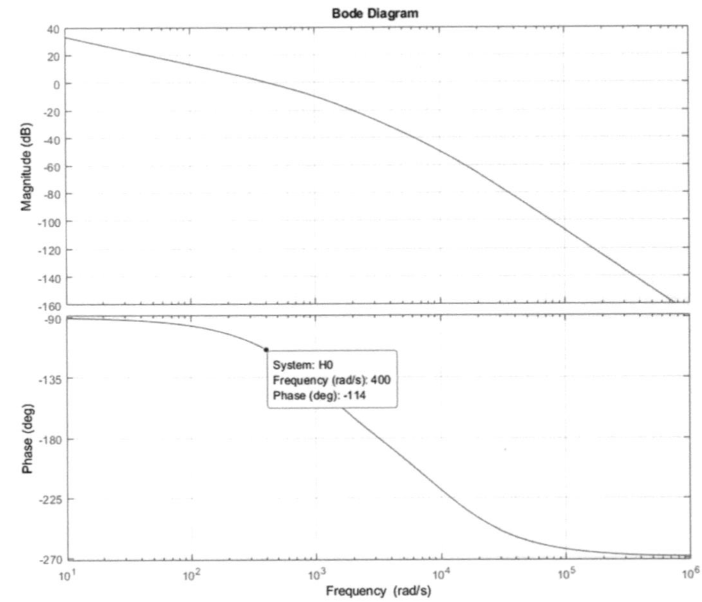

Figure 13.
Bode diagram for an optimal design case.

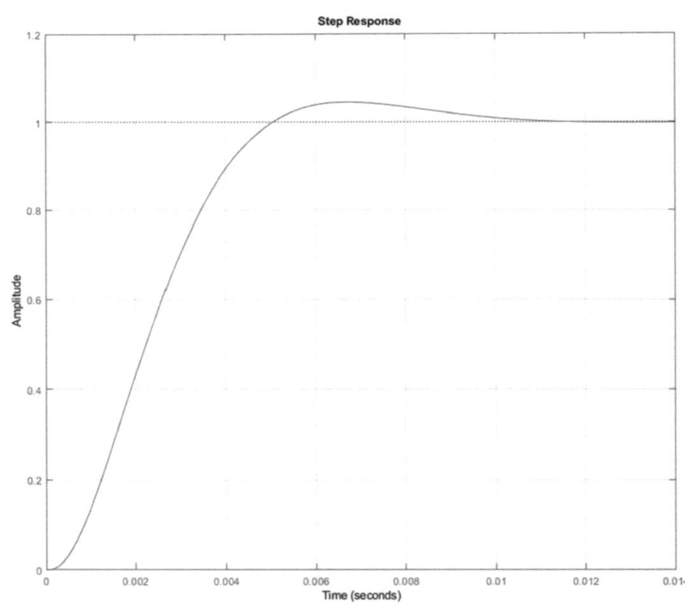

Figure 14.
Step response for an optimal case design.

The phase reserve is around 2°, the control system is stable, but closed to the stability limit.

The response characteristic to a unitary step signal is shown in **Figure 16**.

The response has high oscillations.

In the third case, the controller parameters are chosen out of the stability domain:

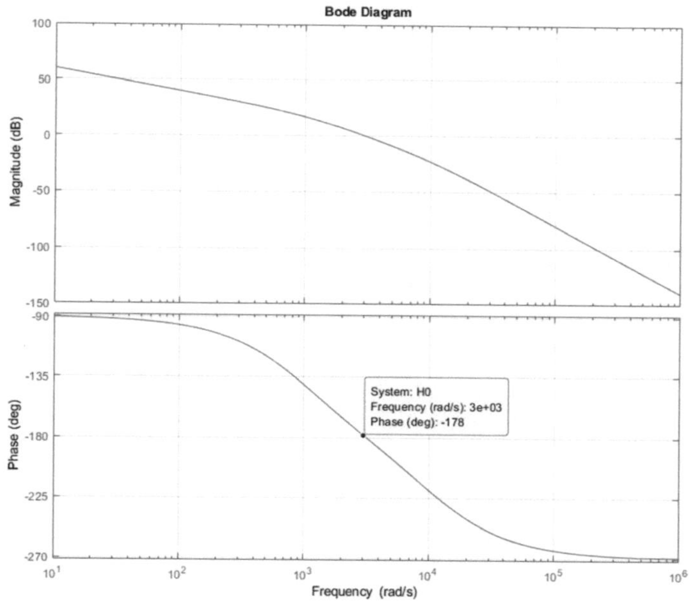

Figure 15.
Bode diagram for parameters closed to the limit of the stability domain.

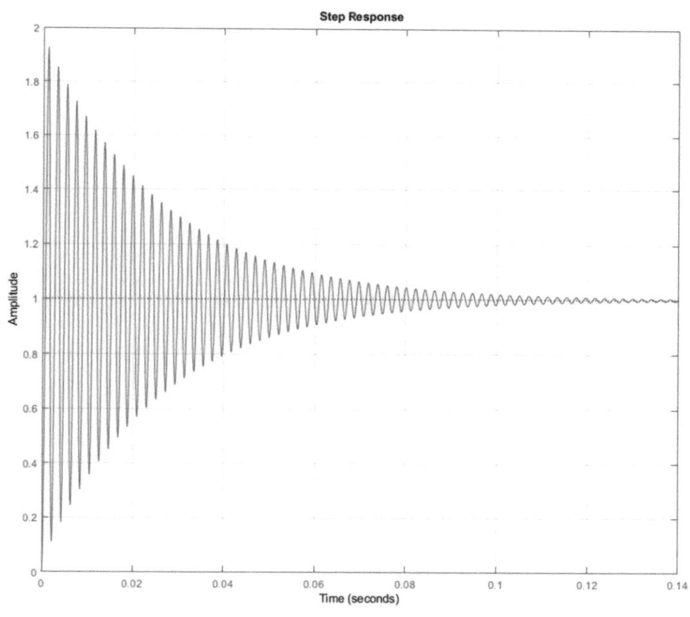

Figure 16.
Step response for parameters close to the limit of the stability domain.

$$T_R = T_1 = 10\ ms, \quad K_R = 125 \tag{39}$$

The amplitude and frequency characteristics are shown in **Figure 17**.
The phase reserve is $-3°$. The system is unstable.
The response characteristic to a unitary step signal is shown in **Figure 18**.
The response is oscillating exponential.

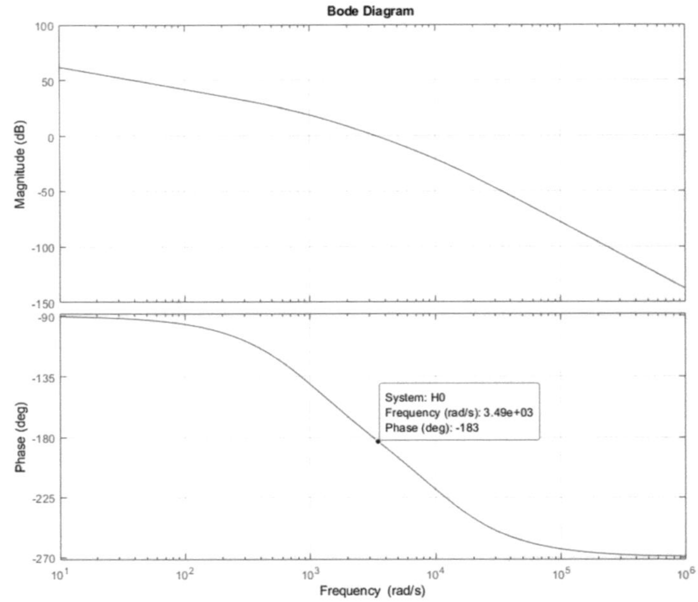

Figure 17.
Bode diagram for parameters out of the stability domain.

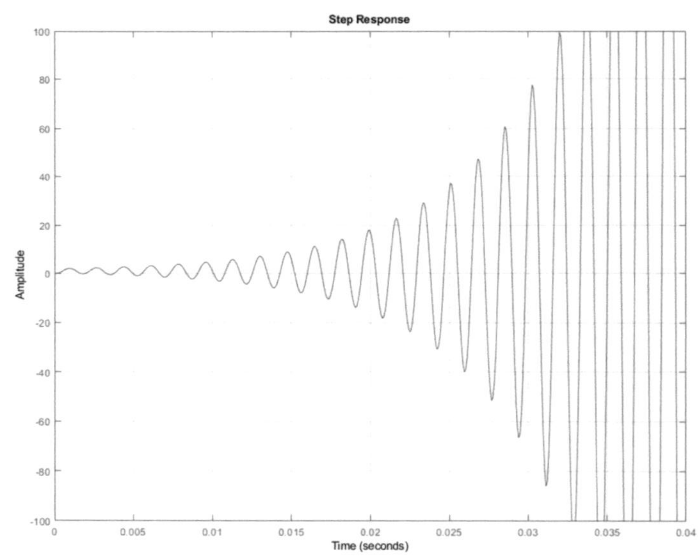

Figure 18.
Step response for parameters out of the stability domain.

4. Discussion

The stability domain for the PI controller parameters in the case of controlling a second order process is represented in two ways in the plane (K_R, T_R) and (T_R, K_R). The stability domains shown in Section 2 are obviously similar. But they were obtained by making two different calculations. The two figures were calculated and plotted for easier analysis. They swapped the abscissa and ordinate. The discussions that can be made for one figure are also valid for the other. It can be seen that T_R can have values smaller positive than a_2/a_1 only for small positive values of K_R. For large values of T_R, the gain K_R may take any real positive value. For large values of K_R, T_R must necessarily take values larger than a_2/a_1. In practice, the time constant of the controller T_R is chosen to be correlated with the time constants of the process T_P. And then, using the phase reserve criterion, the open-loop amplification of the control system is chosen so that the control system ensures a sufficient phase margin, by choosing the controller amplification coefficient K_R.

For the third-order process, the gain K_R can take real positive values only in a finite set, and there is a maximum value for K_{RM}. The relations obvious things namely that the amplification coefficient K_R of the regulator must be smaller than a maximum value, so as to ensure the stability of the closed-loop control system, and the integration time constant T_R of the regulator must be greater that the minimum value, so as to ensure the stability of the closed-loop control system. The time constant T_R can take any value but only in coordination with K_R.

To illustrate the stability domains, several study cases were chosen for each of the two processes analyzed. Thus, a case was chosen in which the controller parameters are outside the stability domain and this was illustrated by the frequency characteristics of the open-loop system, which gave negative phase reserves and by the exponential step signal responses. In other case the parameters were chosen within the stability range, but arbitrarily, without seeking an optimal sizing of the regulator. This led to obtaining frequency characteristics and step signal responses that have low performance indicators. A standard case of study was illustrated where the controller parameters were chosen within the stability domain and calculated using the optimal module criterion in Kessler's version. In this case, it was illustrated that using a designing criterion, the parameters fall within the stability range, and in addition, the performance indicators of the control system have optimal values corresponding to the chosen design criterion.

In theory, the linear regulation system with an PI controller and a second or third order process may be stabilized choosing the controller parameters appropriately depending on the parameters of the controlled process. Knowing the range of values of the stability domain, the controller parameters can be chosen with a tolerance that ensures the stability of the control system for a large variation of the process parameters, in the case of process identification errors or in the case where the process parameters vary over time. The choice of the regulator parameters is however made following a compromise between ensuring a wide margin of stability and ensuring optimal values of the regulation efficiency indicators. The controller parameters are chosen such that the Hurwitz stability domain, which refers to the region of parameters of a system that ensures stability such that its output remains bounded. If the controller parameter values are chosen within the ranges shown, then the system is internally absolutely stable. And from the response to the input signal, it can be observed that the control system is also externally stable in the bounded input-bounded output (BIBO) sense.

5. Conclusion

The stability domains on which the calculations were made in this chapter fall within the broad theory of stability of control systems with application to the control of linear second and third order processes with PI controllers. The control structures with PI controllers are widely used in theory and practice. Their stability can be analyzed using the Hurwitz criterion using mathematical models in continuous time. Mathematical models have standard forms. The study presents a synthesis of the stability domains of PI controller parameters in the case of second- and third-order process control systems and analyzes the specific choice of these parameters. The chapter has a didactic, illustrative character, presenting the domains of stability analytically and graphically, with case studies, simulations and examples. Continuous time controllers may be designed using continuous time criteria. Proper choice of the controller parameters is desirable in practice. The chapter gives guidance for choosing the parameters in the general cases of second and third order processes. The highlighted relationships allow the analytical description of the stability domain, and based on them, this domain can also be illustrated graphically, allowing the designer to easily analyze the mathematical domain in which the values of the controller parameters can be chosen. The increase in the amplification coefficient must be performed in correlation with the modification of the regulator's time constant so that their values remain in the stability range. Several case studies are presented to illustrate how the open-loop and closed-loop control system behaves for various values of the controller parameters chosen in various regions of the stability domain and for domain verification. The presented calculations constitute a theoretical basis, with a teaching role, in the design of linear control systems. The study can be extended to processes of order 4, 5 and so on.

A. Appendix

The appendix presents the MATLAB calculation and simulation program for the second order process.

```
% Controller parameters
KR=5;
TR=0.1e-3;
% Process parameters
KP=1;
T1=10e-3;
T2=1e-3;
a2=T1*T2;
a1=T1+T2;
% Closed-loop system parameters
b1=KR*KP*TR;
b0=KR*KP;
a3n=TR*a2;
a2n=a1*TR;
a1n=TR*(1+KR*KP);
a0n=KP*KR;
% Transfer function of the open-loop system
H0=tf([KP*KR*TR KR*KP],[a2*TR a1*TR TR 0]);
% Transfer function of the closed-loop system
H=tf([b1 b0],[a3n a2n a1n a0n]);
% Plotting the frequency characteristics of the open-loop system
bodeplot(H0)
% Plotting the response to a step signal of the closed-loop system
stepplot(H)
```

Author details

Constantin Volosencu
"Politehnica" University, Timisoara, Romania

*Address all correspondence to: constantin.volosencu@aut.upt.ro

References

[1] Voloşencu C, Saghafinia A, Du X, Chakrabarty S, editors. Control Theory in Engineering. London, UK: IntechOpen Ltd.; 2020. ISBN: 978-1-83880-424-4

[2] Shamsuzzoha M. PID Control for Industrial Processes. London: InTech; 2018. 218 p. DOI: 10.5772/intechopen.69592

[3] Vagia M. PID Controller Design Approaches - Theory, Tuning and Application to Frontier Areas. London: InTech; 2012. 300 p. DOI: 10.5772/2628

[4] Orrante-Sakanassi J, Santibanez V, Camp R. On saturated PID controllers for industrial robots: The PA10 robot arm as case of study. In: Ehsan S, editor. Advanced Strategies for Robot Manipulators. Rijeka: Sciyo; 2010. DOI: 10.5772/10196

[5] Voloşencu C, editor. PID Control - New Design Methods and Applications. London UK: Intechopen; 2025. p. 1003415. DOI: 10.5772/intechopen

[6] Volosencu C. Identification of distributed parameter systems based on sensor networks. In: Meng JE, editor. New Trends in Technologies: Control, Management, Computational Intelligence and Network Systems. Rijeka: Sciyo; 2010. DOI: 10.5772/10407

[7] Han S, Bhattacharyya SP. PID controller synthesis using a σ-Hurwitz stability criterion. IEEE Control Systems Letters. 2018;**99**:1. DOI: 10.1109/LCSYS.2018.2842784

[8] Roskilly T, Mikalsen R. Chapter five - closed-loop stability, marine systems identification. Modeling and Control. 2015:97-122. DOI: 10.1016/B978-0-08-099996-8.00005-7

[9] Haber A. Application of Routh Stability Criterion to Select Parameters of Proportional Integral (PI) Controllers. 2021. Available from: https://aleksandarhaber.com/application –of –routh -stability-criterion-to-select-parameters-of-a-pi-controller/ [Accessed: March 10, 2025]

[10] Abbass AA. Lecture 11 Routh - Hurwitz criterion: Control examples, in ME451: Control Systems. Available from: https://www.academia.edu/123272666/ME451_L11_RouthHurwitzEx [Accessed: March 10, 2025]

[11] Voloşencu C, editor. MATLAB Applications in Engineering. London: IntechOpen; 2022. 118 p. DOI: 10.5772/intechopen.91588